CONCISE ETYMOLOGICAL DICTIONARY OF CHEMISTRY

CONCISE ETYMOLOGICAL DICTIONARY OF CHEMISTRY

STANLEY C. BEVAN, B.Sc., Ph.D., F.R.I.C.

Head of Department of Polymer Science and Technology,
Brunel University, Uxbridge, Middlesex, United Kingdom

S. JOHN GREGG, D.Sc., F.R.I.C.

Associate Senior Research Fellow in Chemistry,
Brunel University, Uxbridge, Middlesex, United Kingdom

ANGELA ROSSEINSKY, M.A.

Sometime Lecturer in Classics,
University of Witwatersrand, Johannesburg, South Africa

APPLIED SCIENCE PUBLISHERS LTD

LONDON

APPLIED SCIENCE PUBLISHERS LTD
RIPPLE ROAD, BARKING, ESSEX, ENGLAND

ISBN 0 85334 653 4

Composed by ETA Services (Typesetters), Ltd., Beccles, Suffolk, England
Printed in Great Britain by Galliard Limited, Great Yarmouth, Norfolk, England

I am not yet so lost in lexicography as to forget that words are the daughters of earth, and that things are the sons of heaven. Language is only the instrument of science, and words are but the signs of ideas. I wish, however, that the instrument might be less apt to decay, and that signs might be permanent, like the things which they denote.

SAMUEL JOHNSON
Preface: *Dictionary of the English Language*

WORKS CONSULTED

The Oxford English Dictionary (Oxford: The Clarendon Press, 1933)

The Oxford Dictionary of English Etymology (Oxford: The Clarendon Press, 1966)

A Comprehensive Etymological Dictionary of the English Language, Ernest Klein (Amsterdam: Elsevier, 1971)

A History of Chemistry, J. R. Partington, Vols. 1–4 (London: Macmillan, 1970)

Encyclopaedia of Chemical Technology, R. E. Kirk and D. F. Othmer (New York: Interscience, 1947)

Thorpe's Dictionary of Applied Chemistry (London: Longmans Green and Co., 1954)

An Etymological Dictionary of Chemistry and Geology, D. Bailey and K. C. Bailey (London: Arnold, 1929)

A Dictionary of Geology, John Challiner (Cardiff: University of Wales Press, 1973)

Introduction to the Study of Minerals, A. F. Rogers (New York: McGraw-Hill, 1937)

'International Union of Pure and Applied Chemistry (IUPAC) Manual of Symbols and Terminology for Physicochemical Quantities and Units,' *Pure and Applied Chemistry*, **21**, No. 1

The Vocabulary of Science, Lancelot Hogben (London: Heineman, 1969)

The Language of Science, T. H. Savory (London: Andre Deutsch, 1967)

A Latin Dictionary, C. T. Lewis and C. Short (Oxford: Clarendon Press, 1951)

A Greek–English Lexicon, H. G. Liddell and R. Scott (Oxford: Clarendon Press, 1907)

PREFACE

In our years of experience as teachers of chemistry we have become increasingly conscious of the value of a knowledge of the origin of the specialist words of our subject; not only does it assist in memorising the meaning of the terms themselves, but it can enliven the whole subject by imparting to it a human dimension. We therefore came to feel the need for a compilation giving the desired information in a handy form. Finding no such work in print (the first, and seemingly only, etymological dictionary of chemistry was published more than forty years ago) we embarked, in collaboration with a teacher of classics, on the task of producing one ourselves.

In the present volume we have collected together information which is scattered in existing dictionaries, reference works and the chemical literature. From the outset it was clear that the compilation would have to be concise rather than comprehensive, and this imposed the necessity of making a selection from the many thousands of possible candidates. We have attempted to follow the rule that a word should be included only if it is widely used in chemistry or if its etymology is particularly interesting or instructive to chemists; but our choice must inevitably reflect in some degree the particular interests, prejudices and tastes of the authors, and will to that extent be arbitrary.

The 'Concise Dictionary' should appeal to teachers of Chemistry in polytechnics and universities, and to practising chemists whether academic or industrial; it should also be of interest and assistance to chemists whose native tongue is other than English, but who need to read and write English in the course of their scientific work. It may perhaps also play a modest part in helping to raise the standard of chemical writing by deepening the interests of chemists in that most important of all their tools, the language of their subject. Lastly, comments from non-scientific friends encourage us to believe that the general reader may find something of interest in its pages. It is perhaps scarcely necessary to add that the emphasis throughout has been on the origin of words rather than on definitions or explanations; the full significance of chemical terms can be conveyed only by a study of systematic treatments in textbooks of the subject.

We gladly acknowledge our indebtedness to the various books, listed separately, which we have consulted, and in particular to *The Oxford*

English Dictionary and Klein's *Comprehensive Etymological Dictionary of the English Language*, without which our task could not have been accomplished. We cordially offer our thanks to Dr B. Ridge for his advice on organic formulae; to Miss Jean Ridge, Mrs Doreen Wray and Mrs Jean Hodges for cheerfully undertaking the typing and to Mr P. Hornsby for assistance in reading the proofs.

CONTENTS

Works Consulted .... *vi*

Preface .... *vii*

Introduction (Including Note on Use) 1

List of Abbreviations 15

Dictionary 17

Appendix 1 – Greek Alphabet 131

Appendix 2 – Latin and Greek Numerals 132

Appendix 3 – Latin Prefixes 133

Appendix 4 – Greek Prefixes 134

Appendix 5 – Latin Roots 135

Appendix 6 – Greek Roots 137

Appendix 7 – Formation of Plurals............. 139

Appendix 8 – SI Prefixes 140

INTRODUCTION

WORDS IN GENERAL

'Then sir, she should have a supercilious knowledge in accounts; and as she grew up, I would have her instructed in geometry, that she might know something of the contagious countries; but above all, Sir Anthony, she should be mistress of orthodoxy, that she might not mis-spell and mis-pronounce words so shamefully as girls usually do; and likewise that she might reprehend the true meaning of what she is saying. This, Sir Anthony, is what I would have a woman know; and I don't think there is a superstitious article in it.'

Mrs Malaprop
in *The Rivals*, by R. B. Sheridan

Words are of inestimable importance to every one of us: we all use them in great number every day for giving and receiving information, as well as for thinking and reasoning. So common are words that we often fail to give them the consideration they deserve. Our power to comprehend and manipulate words can exert a considerable influence on our efficiency and our general happiness, so we ought to make it our business to understand them as fully as possible and to use them correctly. In our professional lives our usefulness depends to a great extent on our ability to communicate with others. Our leisure too can be favourably influenced by a good understanding of the full meaning of words.

In many cases the full significance of words is made clear to us only when we provide ourselves with some knowledge of their origins. But etymology, the branch of linguistics concerned with the origin, formation and development of words, is not commonly taught in schools, and Latin and Greek, so important as the basis of many of the words of modern science, receive far less attention in schools than they did in former times. Yet the need of a good working vocabulary, especially of science, has never been greater than in our own time, when science and technology are transforming the world we live in, at a higher rate than ever before.

The extremely rapid progress of science in recent years has given rise to the creation of many new words, and this expansion of vocabulary has doubtless contributed heavily to the increasing difficulty experienced by the scientist and the non-scientist in their endeavour to attain mutual intelligibility. Non-scientists share with scientists the obligation to improve mutual understanding and to close the gap between the two cultures. Some non-scientists have taken the view that science is in some way the natural enemy of language, and that scientists generally lack conspicuously the ability to communicate with other people. But it could be that the non-scientist, failing to appreciate the logic underlying the vocabulary of science, tends too readily to say, like Lewis Carroll's Eaglet, 'Speak English! I don't know the meaning of half those long words, and, what's more, I don't believe you do either!' And could it be that the scientist too readily takes up an attitude reminiscent of Humpty Dumpty? 'When *I* use a word', Humpty Dumpty said in a rather scornful tone, 'it means just what I choose it to mean – neither more nor less'.

Modern life tends to swamp us all in a flood of words: throughout our waking hours the torrent assails us from books and newspapers, journals and magazines, and from radio and television. Words are an everyday commodity, so plentiful and cheap, and used with so little effort even by the unintelligent and the uneducated, that the handicap of an imperfect or inadequate vocabulary is often not recognised. 'Men imagine', wrote Francis Bacon, 'that their minds have command of language; but it often happens that language bears rule over their minds'. The failing exemplified by Mrs Malaprop is not confined to the rare exception; words are so much part and parcel of our daily lives that it is easy to take them for granted and, through insufficient attention and care, to use them imperfectly and inefficiently. For clear thinking and clear communication, the right words need to be used. The selection of the right words for the task in hand is an important responsibility imposed upon us all.

In one of his letters, Horace observes that although vocational training is universally accepted to be a prerequisite for a career in navigation, medicine, or music, such special preparation is not acknowledged to be a necessary preliminary for the writer of poetry.

> 'A pilot only dares a vessel steer;
> A doubtful drug unlicens'd doctors fear;
> Musicians are to sounds alone confin'd;
> And each mechanic hath his trade assign'd;
> But every desperate blockhead dares to write;
> Verse is the trade of every living wight!'

If this was ever true of the composing of poetry how much more true

2

must it be of the general employment of words: we all use words every day of our lives, but we seldom pause to consider our competence to do so.

The invention of words has been of incalculable consequence to the history of mankind. Without words, civilisation could not have begun, and science and technology could never have been born. Without words, man's life would still be ruled by instinct, like that of beasts: words have emancipated man from an existence limited to the level of primitive urges. Words are precious tools of reasoning and communication, but they are far more than mere implements. In addition to being agents by which the course of history has been guided, they are themselves the product of historical processes, and they enshrine man's cultural history. Like art and architecture, language bears witness to the past, to the world of its creators. Words have an origin and history of their own. They come into existence in response to a need, and when their purpose is served they may change or fade away into disuse, to be replaced by others more suited to the current need.

'As when the forest, with the bending year,
First sheds the leaves which earliest appear,
So an old age of words maturely dies,
Others new-born in youth and vigour rise.'

Many words undergo a striking change of meaning in the course of their lives: when King James II observed that the new St. Paul's Cathedral was amusing, awful and artificial, he was paying Sir Christopher Wren a compliment by declaring his opinion that the building was aesthetically satisfying, awe-inspiring and evidently the product of great skill. The effective use of words is at least as important in the world of today as it was in antiquity. In ancient Greece and Rome, children were trained in the use of their native tongue by means of the works of the greatest poets: thus they learnt, among other things, correctness in the use of words, and elegance of style. In medieval Europe, too, great importance was attached to language, and to proficiency in the use of words. The *trivium* was essentially an education in expression based on studies in Latin, and consisted of *Grammar*, the science of language, involving the study of Vergil, Juvenal, Terence and others, *Rhetoric*, the art of literary composition, learned by study of Quintilian, Cicero and Livy, and *Dialectic*, the development of logical reasoning, based on Boethius and Porphyry. After the trivium came the *quadrivium* which, through the study of geometry, arithmetic, astronomy and music, aimed to give an encyclopaedic knowledge of the world. The prominent place held by Latin and Greek in the schools and universities of western

3

Europe had a profound and far-reaching influence on the development of the vocabulary of science over a period of several centuries. The fact that educated people had a knowledge of Latin and Greek made it a natural practice for new words to be made by direct importing of Latin and Greek words into English, or for new words to be invented from Latin and Greek roots. Only in fairly recent times has the situation arisen in which the users of these words have little acquaintance with the language providing the great majority of the roots from which the words have been constructed.

Words which are the names of things have a special interest and importance, and the manner in which they came into existence was the subject of much speculation in very early times. Plato put forward the view that words were the invention of a single individual, a Lawgiver. A somewhat similar view is recorded in the Bible where, in Genesis, we read that 'God formed every beast of the field and fowl of the air, and brought them to Adam to see what he would call them: and whatsoever Adam called every living creature that was the name thereof. And Adam gave names to all cattle, and to the fowl of the air, and to every beast of the field'.

Lucretius, in his poem of Epicurean Philosophy, ridiculed Plato's theory and, after attributing to natural processes our power of framing sounds, he went on to explain the origin of names as arising by way of mere practical convenience.

'That One the various Names of Things contriv'd,
And that from Him their Knowledge All deriv'd,
'Tis fond to think: for how could that man tell
The Names of Things, or lisp a Syllable,
And not another man perform't as well?
Besides, if others us'd not words as soon,
How was their use, and how the profit known?
Or how could he instruct the Other's mind,
How make them understand what was designed?
For his, being single, neither force nor wit
Could conquer many men, nor they submit
To learn his words, and practise what was fit,
How he persuade those so unfit to hear?
Or how could savage They with patience bear
Strange sounds and words, still rattling in their ear?
But now since Organs fit, since Voice, and Tongue,
By Nature's gift bestow'd, to Man belong,
What wonder is it then, that Man should frame,
And give each different thing a different Name?

4

Since Beasts themselves do make a different noise,
Opprest by pains and fears, or fill'd with joys . . .
How easie was it then for Men to frame,
And give each different Thing a different Name?'

Samuel Johnson, in his *Dictionary of the English Language*, humorously defined 'Lexicographer' as 'a harmless drudge, that busies himself in tracing the original, and detailing the significance of words'. He was very well aware of the great benefit to the progress of science of a good understanding of words, and if we are to realise Johnson's wish that language might be preserved as the instrument of science, we must strive to gain the best possible understanding of the words we use. Fortunately for us, the task of equipping ourselves with a good knowledge of the origin and history of words, far from being drudgery, is often anything but uninteresting or tedious. We are much more likely to find ourselves sharing the excitement and fascination of Cowper's

'Philologists who chase
A panting syllable through time and space,
Start it at home, and hunt it in the dark,
To Gaul, to Greece, and into Noah's Ark.'

By way of example we may quote the interesting and surely unexpected connection between the symbol Ag, for the element silver, and the verb 'to argue'. Both are derived from the same Greek word ἀργός (argos) meaning 'shining, bright or white'. This Greek word yielded the Latin verb 'arguere' which originally meant 'to make as white as silver', and later 'to make clear' or 'to prove'. From the Latin the word entered French as 'arguer' and thence it came into English as 'argue', with its present meaning. Also from ἀργός (argos) came the Greek word ἄργυρος (argyros) signifying 'the white, shining metal', i.e. silver, and the Latin word 'argentum', meaning 'silver' or 'money'. This Latin name for the chemical element silver gave rise to the adoption of the symbol Ag for it in 1813 by Berzelius. Also from argentum came the French word 'argent' meaning 'silver' or 'money', and the English word 'argent' still used in heraldry.

Closely related to ἄργυρος (argyros) is the word ὑδράργυρος (hydrargyros), literally 'water-silver'. This entered Latin as hydrargyrus, which in modern Latin became hydrargyrum, from which came the symbol, Hg, of the chemical element mercury.

Other words stemming from the same Greek root are: (a) litharge, the name of lead monoxide, PbO, from the Greek λίθος, stone, and ἄργυρος, silver, indicating the fact that litharge was obtained as a by-

5

product of the separation of silver from lead, and; (b) the English word 'argil', meaning 'clay', especially white potter's clay, and 'argillaceous' meaning 'largely composed of clay', which reached us through the Greek ἄργιλλος (argillos), white clay, then the Latin 'argilla' and the French 'argille', later 'argile'.

The name Argus, of the giant with a hundred eyes, in Greek mythology also comes from the root ἀργός (argos) meaning 'bright', and can be taken to mean 'the bright one', in allusion to his hundred eyes. Like many words in our tongue, some Greek words have a second meaning. The Greek word ἀργός (argos) also means 'swift' or 'speedy', and it was in this sense that it provided the name Argo for the ship which carried Jason and his fifty-four companions (the Argonauts) to Colchis to recover the golden fleece.

The element argon, symbol Ar, one of the family of noble gases, derives its name from a different Greek word ἀργόν (argon) meaning 'inert' or 'idle'. It was given this name by its discoverers Ramsay and Travers in 1894 because of its outstanding chemical unreactivity. This Greek word ἀργόν (argon) is a form of another word ἀργός (argos) which is a contraction of the word ἀεργός (aergos) literally 'not active', from α- (a-) meaning 'not' and ἔργον, (ergon) meaning work. Since the words 'erg' and 'energy' are derived from ἔργον, we see that they are closely related to the word 'argon'. Another close relative is the word 'metallurgy' which is descended, through the Latin 'metallurgia', from the Greek μεταλλουργός (metalloyrgos) meaning 'metal worker', a word compounded of μέταλλον (metallon) 'metal', and ἔργον (ergon), 'work'.

The process of searching out the origins of words brings as one of its rewards the discovery of many fascinating links of this kind. It provides a hobby having advantages over many other recreations. It helps towards a better understanding of words in general, it improves our acquaintance with scientific technology, thus aiding our comprehension of science, and it discloses a multiplicity of cross-links or bridges between science and language, literature and history.

THE VOCABULARY OF SCIENCE

'If languages are really instruments fashioned by men to facilitate their thinking, they ought to be of the best kind possible; and to try to perfect them is actually to work for the advancement of science.'

Lavoisier

The vocabulary of science has evolved gradually over many centuries. The discovery of new phenomena, the preparation of new substances, the invention of new processes and new instruments, the conception and elaboration of new theories, all these have made necessary the invention of new words. Thus the vocabulary of science has been expanding and changing continuously in parallel with the growth of science itself, and new words have been arising at all stages of the evolution of science.

Up to the thirteenth century, Latin was the only written language of Europe: Magna Carta, for example, was in Latin. For centuries after this, Latin remained the basis of education, and up to the time of Newton (1642–1727) it was the language in which all important books were written, for example: Copernicus, *De Revolutionibus Orbium Coelestium*, 1543; Vesalius, *De Fabrica Humani Corporis*, 1543; Kepler, *Mysterium Cosmographicum*, 1596; Gilbert, *De Magnete*, 1600; Harvey, *De Moto Sanguinis*, 1628; Newton, *Philosophiae Naturalis Principia Mathematica*, 1687; Linnaeus, *Systema Naturae*, 1735.

Some authors used both Latin and English for their writings, e.g. Roger Ascham (tutor to Queen Elizabeth I), Sir Thomas More (beheaded in Henry VIII's time) and Sir Isaac Newton (1642–1727). The Philosophical Transactions of the Royal Society accepted from the beginning (1662) papers in both Latin and English. One of the consequences of the Reformation was that Latin came increasingly to be abandoned as the international language of men of science. The last chemist of renown to use Latin in his writings was the Dutch scientist Boerhaave (1668–1738).

One of the earliest works of a scientific character to be published in English dates from the early days of the English language itself. Not long after the declaration by Edward III in 1362 that henceforth English was to be the speech of Parliament, Geoffrey Chaucer produced for his young son Lowis (i.e. Louis or Lewis) a little book *A Treatise on the Astrolabe*. In this book, published in 1391, Chaucer used several words of foreign origin and explained their meanings. Thus it came about that the words *azimuth*, *nadir* and *zenith* were imported from the Arabic and came to be established in the English Language with the meanings they have continued to possess to this day. Since all educated people of this time were well acquainted with Latin it was quite natural for early writers in English to borrow words from Latin when new terms seemed to be required – as Chaucer did when he introduced the words *equator*, *equinox*, *degree* and *minute* into English. This practice of importing Latin words went on during the following century or two, but during this time the growth of science was proceeding exceedingly slowly, and it is only in the seventeenth century that we can discern what we might call a clearly recognisable vocabulary of science. Following the Reformation the foundations of modern science began to be laid, and the vocabulary of science began to expand more rapidly, a great number of new technical terms being deliberately constructed by the use of Latin and Greek roots. The consequent threat of corruption of the vernacular by the wholesale introduction of foreign words gave rise to concern and in 1664 the Royal Society commissioned one of its founder members, Bishop Wilkins, to inquire into the possibility of creating a universal language for science which would provide a logical and unambiguous medium of communication and at the same time save the common tongue from corruption. This attempt to create an auxiliary medium of communication failed, and scientists turned their attention to the problem of improving the vocabulary of science. Particularly important and influential at this time was the work of Linnaeus (1707–1778) who, in his *Systema Naturae* (1735) introduced a system of biological classification in which Greek roots were freely employed in the construction of new words. From this time on the flow of new words from Latin and Greek roots became a flood, and the vocabulary of science expanded rapidly.

One celebrated example of the adoption of a Greek word for a special scientific purpose is afforded by Stahl's choice, in 1718, of the Greek word φλογιστόν (phlogiston), meaning fire-stuff, for his 'principle of inflammability'. This is of special interest because he gives in detail the reasons for his choice:

'From all these various conditions, therefore, I have believed that it should be given a name, as the first, unique, basic, inflammable principle. But since it cannot, until this hour, be found by itself, outside of all compounds and unions with other materials, and so there are no grounds or basis for giving a descriptive name based on properties, I have felt that it is most fitting to name it from its general action, which it customarily shows in all its compounds. And therefore I have chosen the Greek name phlogiston (signifying inflammable).'

Another famous example, this time from the nineteenth century, is the introduction into the vocabulary of science by Michael Faraday, in 1834, of the terms electrolysis, electrolyte, electrode, anode, cathode, ion, anion and cation. Faraday's investigations of the conduction of electricity by solutions led him to a new theory of the phenomena of electrolysis. His new theory seemed to him to demand a new terminology, and after discussion with William Whewell, classics scholar and later Master of Trinity College, Cambridge, he explained the new terms, as well as the circumstances which had led to their invention:

'The theory which I believe to be a true expression of the facts of

8

electrochemical deposition and which I have therefore detailed in a former series of these Researches, is so much at variance with those previously advanced, that I find the greatest difficulty in stating results, whilst limited to the use of terms which are current with a certain accepted meaning . . . To avoid, therefore, confusion and circumlocution, and for the sake of greater precision of expression than I can otherwise obtain, I have deliberately considered the subject with two friends, and with their assistance and concurrence in framing them, I purpose henceforward using certain other terms, which I will now define . . .'

(Incidentally, Faraday's new theory and new terminology were hotly criticised and rejected by Berzelius on the ground that they were quite unnecessary and unjustifiable.)

Faraday's words 'To avoid confusion and circumlocution, and for the sake of greater precision of expression' convey exactly and concisely the reasons for the growth over the centuries of a special vocabulary of science. Each step forward in the progress of science has made necessary an expansion of the scientific vocabulary, and each expansion of vocabulary has provided science with a more precise and powerful tool for furthering the development of science itself.

Science is far from being unique in having acquired, in the course of time, a special vocabulary serving 'to avoid confusion and circumlocution, and for the sake of greater precision of expression'. Flower lovers, for example, have no difficulty in accepting the Greek or Latin names anemone, azalea, camomile, campanula, chrysanthemum, geranium, gladiolus, hydrangea, iris, phlox, polyanthus, pyracantha, pyrethrum or rhododendron. All specialised activities tend to acquire special sets of words. Enthusiasts for cooking, gardening, sailing, football or cricket, tennis or golf take in their stride the appropriate special terminology. But terms like baste, fillet, roux, corm, tuber, sheet, shroud, spinnaker, penalty, touch, off-side, slip, bye, duck, ashes, hat-trick, maiden over, set, let, serve, birdie, eagle, tee, inevitably offer something of a barrier to understanding by the non-specialist. To the uninitiated, the obstacle is as serious as if the specialisation involved a foreign tongue which had to be mastered first.

In the case of science, the high rate of growth of the subject and the mutually accelerating interaction of science with its vocabulary have brought about the situation that there now exists not only a greater corpus of scientific knowledge than ever before, but also an accumulation of technical terms which appears to most non-scientists to be a formidable barrier to an understanding of science. Of great help in overcoming this barrier is, first, an acquaintance with the fairly limited number of Greek and Latin roots employed in scientific words and, second, some understanding of the logical way, for the most part, in which, over centuries, these words have been constructed in order to meet special needs.

In some cases we are assisted towards an understanding of technical terms by observing relationships in form and meaning between words, both in the field of science and in everyday English, descended from a common root in Latin or Greek. Thus, students tend to mis-spell the name of the first member of the halogen family of elements, transposing the 'o' and 'u' so that the name 'fluorine' wrongly appears to be related to the word 'flour'. These students would be helped by observing the fact that there are many terms in science and in everyday English derived from the Latin word 'fluere', meaning 'to flow', and the associated words 'fluvius' (river) and 'fluctus' (wave). Here are some of them:

fluid: having the property of flowing (as liquids and gases);
fluidised bed: bed of fine particles rendered fluid by agitation;
fluidity: state of being fluid;
flux: a continuing flowing motion (e.g. magnetic flux, neutron flux); a substance used to promote fusion in metal extraction and metal working;
reflux: flowing back again (as in distillation);
influx: inflow;
efflux: outflow;
fluxions: Newton's calculus, concerned with the rate of change of a continuously varying (i.e. flowing) quantity;
fluent: having the gift of flowing speech (also fluency);
effluent: that which flows out, e.g. waste water;
affluent: possessing in flowing abundance (also affluence);
confluent: flowing together;
confluence: junction of two rivers (hence German place-name Coblenz = confluence, where Rhine and Mosel join);
influence: in mediaeval times, the supposed flow from the stars of an ethereal fluid affecting the lives of men (also 'influential');
influenza: illness resulting from supposed effect 'flowing' from the stars (also 'flu');
fluctuation: wave-like motion (also 'fluctuate');
fluorspar or **fluorite:** mineral so called because of its ready fusibility;
fluorine: name coined by Davy for the non-metallic element in fluorspar;
fluoride: compound of fluorine with metal or radical;
fluoro-: prefix meaning fluorine-containing;

10

fluorescence: coloured luminosity produced in certain materials by action of light, so called because the phenomenon was first observed in fluorspar;

fluoresce, fluorescent: see previous word;

fluorene: coal tar hydrocarbon $C_{13}H_{10}$ so called because its impure form fluoresces;

fluorescein: compound formed from resorcinol and phthalic anhydride so named because of its fluorescence;

fluvial: of or appertaining to a river.

Strange as it may seem, the word 'flow' although so similar in form and meaning to all these words, has no etymological connexion with any of them: it is not derived from any Latin root, but comes to us from Old English. Pitfalls like this are not uncommon in etymology.

THE VOCABULARY OF CHEMISTRY

'To call a body which is not known to contain oxygen, and which cannot contain muriatic acid, oxymuriatic acid, is contrary to the principles of that nomenclature in which it is adopted; and an alteration of it seems necessary to assist the progress of discussion, and to diffuse just ideas on the subject. If the great discoverer of this substance had signified it by any simple name, it would have been proper to have referred to it; but 'dephlogisticated marine acid' is a term which can hardly be adopted in the present advanced area of the science.

After consulting some of the most eminent chemical philosophers in the country, it has been judged most proper to suggest a name founded upon one of its most obvious and characteristic properties – its colour, and to call it Chlorine, or Chlorine gas.'

Humphrey Davey: *Philosophical Transactions*, 1811
'Some Reflections on the
Nomenclature of the Oxymuriatic Compounds'

'Hark thou, Anna, thou mayest speak no more of oxymuriatic acid, but may say chlorine: that is better.'

J. J. Berzelius
quoted J. Read: *Humour and Humanism in Chemistry*

The vocabulary of chemistry has evolved over a period of many centuries. Some of the names used in chemistry today were familiar to the alchemists, in the days long before chemistry existed as an exact science, before the name chemistry had been invented. Thus we find some familiar current chemical names mentioned by Chaucer, writing

11

in the fourteenth century concerning the supposed relation between the seven metals and the heavenly bodies:

'The bodies seven, eek, lo heer anon
Sol gold is, and Luna silver we declare;
Mars yron, Mercurie is quyksilver;
Saturnus leed, and Jubiter is tyn,
And Venus coper, by my fathers kyn.'

and again when he lists some of the materials used in the alchemists' laboratory:

'Arsenek, sal armoniak, and bromstoon,
And herbes could I tell eek many a one,
As egremoigne, valerian, and lunarie,
And other suche, if that me list to tarie . . .
Unslekked lyme, chalk and glare of an ey,
Cered poketts, sal petre, vitriole;
And dyvers fyres made of woode and cole;
Salt tartre, alcaly, and salt preparat,
And combust materes, and coagulat;
Cley made with hors or mannes hair, and oyle,
Of tartre, alym, glas, barm, wort and argoyle.'

In the Authorised Version of the Bible, published in 1611, we meet the names of gold, silver, iron, lead and tin, also brass (copper) and brimstone (sulphur).

The complexity of the terminology of alchemy is well illustrated in the following extract from the play Gallathea, published in 1632 by John Lilley. In this conversation between Peter, the alchemist's young assistant, and Rafe, we note that the list of common laboratory materials bears a striking resemblance to the one given by Chaucer two and a half centuries earlier:

Peter: What a life doe I lead with my Master, nothing but blowing of bellowes, heating of spirits, and scraping of Croslets? it is a very secret Science, for none almost can understand the language of it, Sublimation, Almigation, Calcination, Rubification, Encorporation, Circination, Sementation, Albification and Frementation. With as many terms unpossible to be uttered, as the Arte to bee compassed.
Rafe: Let me crosse my selfe, I never heard so many great devils in a little Monkies mouth.

12

Peter: Then our instruments, Croslets, Sublivatories, Cucurbits, Limbecks, Decensores, Violes, manuall and murall, for embibing and conbibing, Bellowes, molificative and endurative.
Rafe: What language is this? doe they speak so?
Peter: Then our Metals, Saltpeeter, Sal Armonick, Egrimony, Lumany, Brimstone, Tartar, Alam, Breem Worte, Glasse, Unsleked lime, Chalke, Ashes, hayre and what not, to make I know not what.
Rafe: My haire beginneth to stand upright, would the boy make no end.

The transformation of alchemy into chemistry in the seventeenth and eighteenth centuries, and the adoption, followed by the abandonment, of the phlogiston theory, imposed tremendous strains on the language, and towards the end of the eighteenth century change followed swiftly on the heels of change. One example of this revolution in terminology is quoted at the beginning of this section. Scheele in 1774 had prepared chlorine by the oxidation of hydrochloric acid, and in the language of his time gave it the name 'dephlogisticated marine acid gas'; Lavoisier's work on oxygen and his recognition of its role as an acid-producing element, together with Berthollet's observation in 1785 that a solution of chlorine in water gave off bubbles of oxygen and left a solution of 'muriatic acid' led to a change of name to 'oxymuriatic acid', but this term was short-lived, and in 1811 Davy felt compelled, as a result of his researches, to change the name again, and invented 'chlorine'.

Another example of this acute instability of terminology is found in text-books of the period in sections dealing with oxygen. In Gregory's *Economy of Nature* (1804) it was deemed necessary to introduce the subject of oxygen with these explanatory paragraphs:

'It has been already intimated that *gas*, signifying spirit or ebullition, was a term employed by Van Helmont, and other Dutch and German chemists, to describe those elastic fluids which appeared in their nature different from common or atmospheric air . . . The fluid under our immediate consideration was originally termed *dephlogisticated* air, a name given it by Dr. Priestley from supposing it free from phlogiston or inflammable matter; when it was found essential to animal life, it obtained the name of *pure* or *vital* air; and when it was known to contribute essentially to ignition, and the other phenomena of fire, it was termed *empyreal* air; but the French chemists, having discovered that its basis is the substance which imparts the acid character to all the mineral and vegetable acids, have distinguished it by the name of *oxygen gas*.'

Thus in the words of T. S. Eliot,

'Words strain,
Crack, and sometimes break, under the burden,
Under the tension, slip, slide, perish,
Decay with imprecision, will not stay in place,
Will not stay still.'

Gregory's reference to the 'French chemists' relates to Lavoisier's researches on the nature of combustion and to the publication in 1787 of the fundamentally important *Methode de Nomenclature Chimique* under the initiative of de Morveau, Lavoisier, Berthollet and Fourcroy. In a memoir to the Academy of Science Lavoisier listed some contemporary chemical names of little value to discussion and to the progress of science (e.g. powder of algaroth, salt of alembroth, turbrith mineral, oil of vitriol, butter of antimony), and in a later publication he put forward a reformed designation for some names of this kind (e.g. copper sulphate for blue vitriol).

Lavoisier displayed a rare insight into the function of language in science. He declared languages to be 'not merely positive signs to express thought' but also 'analytical systems by means of which we advance from the known to the unknown and to a certain extent in the manner of mathematics . . . if languages are really instruments fashioned by men to make thinking easier, they should be of the best kind possible, and to strive to perfect them is indeed to work for the advancement of science . . . this method which must be introduced into the teaching of chemistry is closely connected with the reform of the nomenclature. A well-composed language adapted to the natural and successive order of ideas will bring in its train a necessary and immediate revolution in the method of teaching . . . we shall have three things to distinguish in every physical science: the series of facts that constitute the science, the ideas that call the facts to mind, and the words that express them. The word should give birth to the idea; the idea should depict the fact'.

NOTE ON THE USE OF THE DICTIONARY

Bold type, when used in an entry elsewhere than in the heading, indicates that the word (or part of a word, or symbol) appears as an entry heading in its alphabetical position in the dictionary.
Italics. If a proper name is involved in the etymology of a word, it is printed in italics. Thus '**GADOLINIUM** n After *Johan Gadolin*', but '**MERCAPTAN**, n Coined by Zeise'.

14

LIST OF ABBREVIATIONS

acc	accusative	ME	Middle English
adj	adjective	min	mineralogy, mineralogical
adv	adverb	Mod L	Modern Latin
Arab	Arabic		
		n	noun
c, ca	circa	nom	nominative
cf	compare (*confer*)		
chem	chemistry, chemical	OE	Old English
dim	diminutive	OF	Old French
		OI	Old Indian
e.g.	for example (*exempli gratia*)	ON	Old Norse
Eng	English		
		Pers	Persian
F	French	pl	plural
foll	following	pp	past participle
fr	from	prec	preceding
		prep	preposition
gen	genitive	pres p	present participle
Ger	German	pron	pronounced
Gk	Greek		
		Sp	Spanish
i.e.	that is (*id est*)	suff	suffix
IE	Indo–European		
Ital	Italian	v	verb
		viz	namely (*videlicet*)
L	Latin	voc	vocative
lit	literally		
		wd	word
Med L	Medieval Latin		

A

A Symbol for **ampere** (q.v.), also for area. Up to 1957, chemical symbol for the element argon, but then superseded by symbol **Ar**. Symbol for Helmholtz energy $(U - TS)$.

Å Symbol for **Ångström unit** $(10^{-10}\ m)$.

a Symbol for **atto-** (q.v.), symbol for 10^{-18} in the SI system.

a- From Gk $\dot{a}$- (a-), 'without' (see Gk prefixes Appendix 4).

ABIETIC adj L *abies*, gen *abietis*, 'fir tree'. Abietic acid, $C_{19}H_{28}COOH$, is found in pine resin.

ABSCISSA n L *ab*, 'away from', *scissa*, 'cut'; hence, a distance cut off, i.e. by the perpendicular from a point to the *x*-axis.

ABSORB v L *ab*, 'away from', *sorbere* 'to swallow up'. To imbibe or take up physically or chemically, e.g. the taking up of hydrogen by palladium.

ABSORPTION n See prec wd. The process of taking up a gas or liquid into the body of a solid as distinct from onto the surface of a solid (see **adsorption**).

AC or **ac** Abbrev for alternating current as distinct from direct current dc.

Ac Chemical symbol for **actinium** (q.v.), element at. no. 89.

ACENAPHTHENE n From **ace(tic)** and **naphth(al)ene**. The hydrocarbon, $C_{12}H_{10}$, closely related to naphthalene.

ACET- Combining form of **acetic** (see foll wds).

ACETAL n From **acet(ic)** and **al(cohol)**. The product of condensation of an aldehyde with an alcohol, e.g. CH_3CHO (acetaldehyde) + $2C_2H_5OH$ (ethyl alcohol) = $CH_3CH-(OC_2H_5)_2$ (acetal) + H_2O.

ACETALDEHYDE n From **acet(ic)** and **aldehyde** (q.v.). The volatile liquid CH_3CHO which yields acetic acid on oxidation.

ACETAMIDE n From **acet(yl)** and **amide** (q.v.). The colourless crystalline substance CH_3CONH_2, the amide of acetic acid.

ACETANILIDE n From **acet(yl)** and **anilide.** The acetyl derivative of aniline, $C_6H_5NHCOCH_3$.

ACETATE n From **acet(ic)** and **-ate** (q.v.). Salt or ester of acetic acid.

ACETIC adj L *acetum*, 'vinegar'. Acetic acid, CH_3COOH, the acid of vinegar: crystalline solid at low temperatures (hence the descriptive term 'glacial' applied to the pure substance).

ACETIN n From **acet(ic)** and (glycer)in (old name for **glycerol**). Name of the mono-, di- or tri- acetyl derivatives of glycerol (cf. **palmitin, stearin**).

ACETO- Combining form of **acetic.**

ACETONE n Coined by Bussy in 1833, fr **acet(ic)** and **-one** (q.v.) for dimethylketone $(CH_3)_2CO$, in reference to its preparation by the dry distillation of calcium acetate.

ACETYL adj From **acet(ic)** and **-yl**. Name of the radical CH_3CO.

ACETYLENE n Coined by Berthelot 1864 for the gaseous hydrocarbon C_2H_2, since it was derived from 'acetyl' (then C_2H_3) in the same way as ethylene from ethyl: $C_2H_3 - H = C_2H_2$; $C_2H_5 - H = C_2H_4$.

ACETYLIDE n See prec wd. A metal derivative of acetylene, e.g. copper (I) acetylide, Cu_2C_2, silver (I) acetylide, Ag_2C_2.

ACI- Combining form of **acid**. The acid form of a substance, applied particularly in the case of tautomerism, e.g. nitroethane, normal form $C_2H_5NO_2$, aci-form CH_3CH: NOOH.

ACICULAR adj L *acicula*, 'a small pin, prickle' dim of *acus*, 'needle, pin'. Long and thin, needle-shaped (applied especially to crystals).

ACID n or adj F *acide*, fr L *acidus*, 'sour' (cf. L *acetum*, 'vinegar'). A substance capable of donating protons or accepting electrons.

ACIDIFY v L *acidus* 'sour' and **-fy**, fr L *facere*, 'to make'. To make acid.

ACIDIMETRY n L *acidus* 'sour', Gk μέτρον (metron) 'measure'. The measurement of acidity.

ACIDULATE v L *acidulus*, dim of *acidus*, 'sour'. To make slightly acid.

ACONITIC adj F *aconit*, fr L *aconitum*, fr Gk ἀκόνιτον (aconiton), 'monkshood'. Aconitic acid, the acidic substance found in the poisonous plant monkshood (also known as wolf's-bane).

ACONITINE n From aconite (see prec wd) and suff **-ine**, denoting an alkaloid. An alkaloid found in monkshood.

ACRIDINE n From Eng *acrid* 'sharp, bitter', fr L *acer*, *acris*, 'pungent'. A substance $C_{12}H_9N$ found in coal tar, which when inhaled causes violent sneezing.

ACROLEIN n L *acer*, *acris* 'sharp, bitter' and L *olere*, 'to smell'. Name given by Gmelin to the volatile liquid of pungent odour formed in the thermal decomposition of glycerol; the unsaturated aldehyde, $CH_2: CHCHO$.

ACRYLATE n See foll wd. A salt or ester of acrylic acid, e.g. methyl acrylate, $CH_2: CHCOOCH_3$.

ACRYLIC adj From **acr(olein)**, **-yl** and **-ic**. Acrylic acid, $CH_2CHCOOH$, the unsaturated acid formed on oxidation of acrolein.

ACRYLONITRILE n From **acryl(ic)** (see prec wd) and **nitrile** (q.v.); the nitrile corresponding to acrylic acid. The unsaturated compound $CH_2: CHCN$.

ACTINIC adj Gk ἀκτίς (actis) gen ἀκτῖνος (actinos) 'ray'. Term used to denote that part of the sun's radiation reaching the surface of the earth which is active in initiating or promoting chemical reaction, especially ultraviolet radiation.

ACTINIDE n From **actinium** (q.v.), element 89. Name given to the element actinium (at. no. 89) and all the elements following it in the periodic table, i.e. to all the elements with vacancies in the 5f electron shell.

ACTINIUM n Gk ἀκτίς (actis) gen ἀκτῖνος (actinos) 'ray, radiance'. Name given by Debierne in 1899 to the radioactive element (at. no. 89) extracted from pitchblende.

ACYCLIC adj Gk ἀ-(a-) 'without', κύκλος (cyclos), 'circle'. Term used in reference to compounds in the molecules of which the atoms are not arranged in rings (cf. **cyclic, homocyclic** and **heterocyclic**).

ACYL adj From **ac(id)** and **-yl** (q.v.). Name of the radical of a carboxylic acid, e.g. acetyl, CH_3CO and benzoyl, C_6H_5CO are acyl radicals.

ADENINE n Gk ἀδήν (aden) 'gland'. The purine compound $C_5H_5N_5$ prepared (by Kossel, in 1885) from the nucleotides of the pancreas (the gland near the stomach).

ADIABATIC adj From Gk ἀ-(a-), 'not', διά (dia), 'through', and βαίνειν (bainein), 'to go, to pass'. Term used in reference to processes in which no heat passes into or out of the system.

ADIPIC adj L *adeps*, gen *adipis*, 'fat'. Adipic acid, a dicarboxylic acid $(CH_2)_4(COOH)_2$ first obtained by the oxidation of fats.

ADRENALIN n From L *ad*, 'to', *renalis*, 'pertaining to the kidneys'. The crystalline substance $C_6H_3(OH)_2$-$CH(OH)CH_2NHCH_3$ first prepared by Takamine in 1901 from the adrenal glands (which are near the kidneys).

ADSORB v L *ad*, 'to', *sorbere*, 'to swallow up'. To take up and hold on the surface of a solid, by physical or chemical attraction, either a gas or vapour, or a dissolved substance from the liquid phase.

ADSORPTION n See prec wd. The taking up and holding, on a solid surface, of a gas or vapour, or of a dissolved substance from the liquid phase.

AEROBIC adj Gk ἀήρ (aer) 'air', βίος (bios) 'life'. Name coined by Pasteur in 1863 to denote bacterial processes occurring only in the presence of oxygen (opp anaerobic).

Ag From L *argentum*, 'silver'. Chemical symbol for **silver**, element at. no. 47.

AGAR or **AGAR-AGAR** n Malay word. A gelatinous substance obtained from sea-weed.

AGATE n F, fr L *achates*, fr Gk ἀχάτης (achates) 'agate'. A hard precious stone, a form of silica.

AGLYCONE n From Gk ἀ-(a-), 'without', **glucose** (q.v.). A substance combined with glucose or some other sugar in a compound called a glucoside or glycoside respectively (see, for example, **amygdalin** or **anthocyanidin**).

19

Al Chemical symbol for **aluminium**, element at. no. 13.

-AL Suffix usually denoting an **al(dehyde)**.

ALANINE n From **al(dehyde)**, *-an-* (euphonic insertion) and *-ine* (from amine). The amino acid $CH_3CH(NH_2)COOH$, α-aminopropionic acid. The name derives from the original method of preparation from aldehyde-ammonia by Strecker in 1850.

ALBITE n L *albus*, 'white'. A white felspar, $NaAlSi_3O_8$.

ALCOHOL n Med L, fr Arabic *al-kohl*, the fine dark powder used in the East as eye-shadow; also, any similar fine impalpable powder produced by grinding or by sublimation; hence, by extension, to fluids prepared by distillation. Alcohols are hydroxyl derivatives of hydrocarbons, the best known being ethyl alcohol, or ethanol, C_2H_5OH, and methyl alcohol, or methanol, CH_3OH.

ALDEHYDE n From **al(cohol)** **dehyd(rogenatum)**. Name coined by Liebig in 1837 to denote 'alcohol deprived of hydrogen'. A compound containing the group CHO, e.g. acetaldehyde CH_3CHO.

ALDOL n From **ald(ehyde)** and **(alcoh)ol**. Name coined by Wurtz in 1872 to indicate part-aldehyde and part-alcohol nature. Compounds of the type $CH_3CH(OH)CH_2CHO$, formed by the 'aldol reaction', e.g. $2CH_3CHO \rightarrow CH_3CH(OH)CH_2CHO$.

ALDOSE n From **ald(ehyde)** and *-ose* (suff signifying a sugar). Sugars which are aldehydes, e.g. glucose, $CHO(CHOH)_4CH_2OH$, as distinct from those which are not, e.g. fructose, a ketose, $HOCH_2CO(CHOH)-CH_2OH$.

ALGA (pl algae) n L *alga*, 'seaweed'. Seaweed and certain similar freshwater plants.

ALGINATES n L *alga*, 'seaweed', salts of a polymeric acid, $(C_6H_8O_6)_n$, found in seaweed.

ALICYCLIC adj From **ali(phatic)** and **cyclic**. Term devised by Bamberger in 1889 to denote aliphatic compounds containing closed carbon rings, e.g. cyclohexane, $(CH_2)_6$.

ALIPHATIC adj Gk ἄλειφα, gen ἀλείφατος (aleiphatos) 'fat'. Term introduced by Hjelt, ca. 1860, to distinguish those carbon compounds found in fats from those found in or derived from aromatic substances. Open chain hydrocarbons and their derivatives as distinct from derivatives of benzene.

ALIQUOT n L *aliquot*, 'a few', from L *alius*, 'other', and *quot*, 'how much'. A portion taken for analysis and which is a known fraction of the whole sample.

ALIZARIN n F *alizarine*, fr Sp *alizari*, fr Arabic *al-asara*, 'madder root'. A dye, $C_{14}H_8O_4$ present in madder. The importance of natural madder as a dye declined following the synthesis of alizarin by Graebe and Liebermann in 1869.

ALKALI n F *alcali*, fr Arabic *al qili*, 'the roasted', the product of roasting marine plants, hence, any substance having properties similar to those of sodium carbonate. A water-soluble base yielding a caustic solution.

ALKALOID n From **alkali** and Gk -οειδής (-oeides), 'like', (cf. F *alcaloide*). A hybrid term used by Dumas in 1835 to describe those substances known up to that time by the name 'vegetable alkalis', a description due to Berzelius. Organic nitrogen bases occurring in plants and having powerful action on animals, e.g. nicotine and morphine.

ALKANE n From **alk(yl)** and **(meth)ane**. Member of the paraffin series of hydrocarbons, e.g. methane, ethane, propane.

ALKYL adj From **alk(ali)** and **-yl**. A hydrocarbon radical of the paraffin series, e.g. methyl, CH_3, ethyl, C_2H_5. Name first given to the ethyl radical on account of its ability to form compounds analogous to those of the alkali metals.

ALLANTOIN n Gk ἀλλαντοειδής (allantoeides), 'sausage-shaped'. Name given by Wöhler in 1838 to the crystalline substance, $C_4H_6N_4O_3$, related to uric acid and found in the fluid of the foetal membrane (which acquired its name, the allantois, because of its shape in the calf).

ALLENE n From **all(yl)** and **-ene** (suff denoting an unsaturated hydrocarbon). Unsaturated hydrocarbon containing two adjacent double bonds, e.g. $CH_2:C:CH_2$.

ALLOTROPE n Gk ἄλλος (allos), 'other', τρόπος (tropos), 'turn, direction'. A particular distinct form of an element, thus oxygen and ozone are allotropes (see foll wd).

ALLOTROPY n See prec word. Term introduced by Berzelius in 1840 to denote the existence of an element in different distinct molecular or crystalline forms, e.g. carbon as diamond and graphite.

ALLOXAN n From **all(antoin)** and **oxa(lic)**. Name given by Wöhler and Liebig to the substance $C_4H_2N_2O_4$ because they supposed it to be composed from allantoin and oxalic acid.

ALLOY n F *aloi*, fr L *alligare*, fr L *ad*, 'to', and *ligare*, 'to bind'. An intimate association (which may be a compound, solution or mixture) of two or more metals, which has metallic properties.

ALLYL adj L *allium*, 'garlic'. The hydrocarbon radical $CH_2:CHCH_2$, so called by Wertheim in 1844 because it was found, as the sulphide, in oil of garlic.

ALPHA-RAYS n Gk ἄλφα (alpha), first letter of Gk alphabet. Name given in early investigations of radioactivity (1899) to the less penetrating radiation given off by uranium in radioactive decay and later shown to consist of particles identical with the helium nucleus.

ALUM n L *alumen*, 'bitter salt'. Name given to certain double sulphates which crystallise readily as octahedra of composition $M^IM^{III}(SO_4)_2.12H_2O$ and typified by common alum $KAl(SO_4)_2.12H_2O$.

ALUMINA n From prec wd. Aluminium oxide, Al_2O_3.

ALUMINATE n From prec wds. Salt formed by alumina acting as an acidic oxide.

ALUMINIUM or **ALUMINUM** (U.S.) n See prec wds. From Davy's names (1812) for the light metal present in alumina and alum. Chemical element at. no. 13; symbol **Al**.

Am Chemical symbol for **Americium**, element at. no. 95.

AMALGAM n F *amalgame*, fr Mod L *amalgama*, fr Arabic *al*, 'the', and Gk μάλαγμα (malagma) 'a soft mass'. Originally, a soft alloy of mercury with another metal. Now, any alloy of mercury.

AMATOL n From **am(monium nitr)at(e)** and **(trinitro)tol(uene)**. Name coined in 1918 for the high explosive made from ammonium nitrate and TNT.

AMBIENT adj L *ambiens*, gen *ambientis*, 'surrounding'. Surrounding, as in 'ambient temperature', the temperature of the surroundings.

AMERICIUM n From *America*. Name given by its discoverer Seaborg in 1946, after the two Americas, by analogy with the corresponding lanthanide element europium, at. no. 63, named after Europe. Transuranium element at. no. 95.

AMETHYST n Gk ἀμέθυστος (amethystos), 'cure for drunkenness', name derived from the ancient belief in the power of this stone to prevent intoxication. Precious stone of violet colour, consisting of quartz tinted by traces of impurity.

AMIDE n From **am(monia)** and **-ide** (q.v.); name given by Wurtz to compounds derived from ammonia by replacement of a hydrogen atom by an acid radical, e.g. acetamide, CH_3CONH_2, and benzamide, C_6H_5-$CONH_2$. Also, name of metal derivatives of ammonia, e.g. sodamide $NaNH_2$.

AMINE n From **am(monia)** and **-ine** (q.v.). Name of compounds derived from ammonia by replacement of one or more of its hydrogens by alkyl or other hydrocarbon radicals, e.g. methylamine, CH_3NH_2, ethylenediamine $NH_2CH_2CH_2NH_2$.

AMINO- See prec wd; combining form of **amine**. Containing the amine group; thus aniline, $C_6H_5NH_2$, is aminobenzene.

AMMETER n Abbreviation of ampere-meter (see **ampere**). Instrument for measuring electric current.

AMMINE n From **amm(onia)** and **-ine** (q.v.). Complex in which ammonia is directly linked to the central metal atom.

AMMONIA n From *sal ammoniac*, the salt (ammonium chloride) supposed to have been first made in Armenia or in the vicinity of the shrine of Jupiter Ammon. The gas, NH_3.

AMMONIUM adj From **ammonia** and **-ium** (by analogy with name ending of metals, Berzelius). The positive ion NH_4^+ derived from ammonia.

AMMONO- From **ammonia**. Derived from ammonia; thus in liquid ammonia acetamide is an ammono-acid, and ammono-lysis is the reaction corresponding to hydrolysis in water.

AMORPHOUS adj Gk *ἀ-* (a-) 'without', *μορφή* (morphe) 'shape'. Non-crystalline, a solid lacking regular arrangement of its ions, atoms or molecules.

AMP n Abbreviation of **ampere** (see foll wd).

AMPERE n After *Andre Marie Ampère* (1775–1836). Unit of electric current (SI system).

AMPEROMETRY n From **ampere** (see prec wd) and Gk *μέτρον* (metron) 'measure'. Method of analysis based on measurement of electric current.

AMPHETAMINE n Abbreviated from alpha-methylphenyl ethylamine, $C_6H_5CH_2CH(CH_3)NH_2$, a synthetic drug which stimulates the heart and respiration.

AMPHIBOLE n F, fr L *amphibolus*, 'ambiguous', fr Gk *ἀμφί* (amphi) 'around, on both sides', and *βάλλειν* (ballein), 'to throw'. A group of asbestos minerals so named by Haüy (1743–1822) in reference to their variable composition.

AMPHIPROTIC adj Gk *ἀμφί* (amphi) 'on both sides', and **proton** (q.v.). Both yielding and accepting protons. Term used to denote solvents, e.g. water, which are both protogenic and protophilic.

AMPHOTERIC adj Gk *ἀμφοτέρως* (amphoteros), 'in both ways'. Possessing both acidic and basic properties.

AMYGDALIN n L *amygdala*, fr Gk *ἀμυγδάλη* (amygdale) 'almond'. The glycoside $C_{20}H_{27}NO_{11}$ found in almonds and other stone fruit, and which on hydrolysis yields glucose, benzaldehyde and hydrocyanic acid.

AMYL adj L *amylum* fr Gk *ἄμυλον* (amylon) 'starch'. Name given by Balard in 1844 to the alcohol which had been obtained from starch. The hydrocarbon radical $C_5H_{11}\cdot$.

AN Abbreviation of **ane** (q.v.) used in the penultimate position of a chemical name indicating a derivative of a saturated hydrocarbon, e.g. methanol.

ANACARDIC adj From *Anacardium occidentale*, the name of the cashew-nut tree. Anacardic acid, the acid found in cashew-nut shell liquid.

ANAEROBIC adj Gk *ἀν-* (an-), 'without', *ἀήρ* (aer) 'air', *βίος* (bios) 'life'. Term coined by Pasteur in 1863 to denote bacterial processes occurring in the absence of oxygen (opposite of **aerobic** (q.v.)).

ANAESTHESIA n Gk *ἀν-* (an-) 'without', *αἴσθησις* (aisthesis), 'feeling'. Absence of feeling or sensation.

ANAESTHETIC adj or n See prec wd. Substance producing loss of feeling or sensation.

ANALCITE n Gk *ἀν-* (an-) 'without', *ἀλκή* (alke) 'strength', so called because it acquires only a weak electric charge when rubbed. A hydrated mineral of the zeolite group, $NaAlSi_2O_6 \cdot 6H_2O$.

ANALGESIA n Gk ἀν- (an-) 'without', ἄλγος (algos) 'pain'. Absence of pain, as distinct from loss of feeling (see **anaesthesia**).

ANALGESIC adj or n See prec wd. Pain-relieving drug.

ANALYSE or **ANALYZE** v See full wds. To determine chemical composition.

ANALYSIS (pl **analyses**) n Gk ἀνά (ana) 'up', throughout', λύσις (lysis) 'loosing, setting free'. Separation into constituent parts (opposite of **synthesis**, 'putting together'), hence, determination of chemical composition.

ANALYST n See prec wd. A person engaged in the determination of chemical composition.

ANATASE n F *anatase*, Gk ἀνάτασις (anatasis) fr Gk ἀνά (ana) 'up', τείνειν (teinein) 'to stretch', so named by Haüy in 1843 on account of the length of the crystals of this mineral. A particular form of titanium dioxide (cf. **rutile**).

-ANE L -*anus*, word-ending denoting relationship of some kind. In chemistry, the name-ending proposed by Hofmann in 1866 to denote membership of the paraffin hydrocarbons, e.g. pentane C_5H_{12}.

ÅNGSTRÖM UNIT n After *Ångström* (1814–74). A unit of length, symbol Å (1 Å$= 10^{-10}$ m$= 10^{-1}$ nm).

ANHYDRIDE n Gk ἀν- (an-) 'without', ὕδωρ (hydor) 'water'. A substance derived from another by removal of the elements of water, e.g. acetic anhydride $(CH_3CO)_2O$.

ANHYDROUS adj See prec wd. Deprived of water, e.g. anhydrous copper sulphate, $CuSO_4$, as distinct from the hydrate $CuSO_4.5H_2O$.

ANILIDE n From **aniline**. A derivative of aniline made by replacement of amino-hydrogen atom, e.g. acetanilide $C_6H_5NHCOCH_3$.

ANILINE n Ultimately fr Arabic, *al*, 'the', *nil* 'indigo'. The name given by Fritzsche in 1841 to the liquid substance he obtained by distilling indigo with alkali. The aromatic amine, aminobenzene, $C_6H_5NH_2$.

ANION n Gk ἀνίον (anion) 'something going up', fr Gk ἀνά (ana) 'up' and ἰέναι (ienai) 'to go'. Name given by Faraday in 1834 to the negatively charged ion in electrolysis (see **ion, cation**), because it moved towards the anode.

ANIONIC adj See prec wd. Having the properties of an anion.

ANISOLE n F fr L *anisum*, fr Gk ἄνισον (anison) 'anise, dill', so named on account of its occurrence in essence of anise. The liquid substance p-$C_6H_5OCH_3$.

ANISOTROPIC adj Gk ἀν- (an-) 'not', ἴσος (isos) 'equal', τρόπος (tropos) 'turn'. Possessing different properties in different directions.

ANNULENE n L *annulus* (correctly, *anulus*) 'ring'. Any one of a group of organic compounds containing a ring of carbon atoms joined by alternate single and double

24

bonds, i.e. by conjugated double bonds.

ANODE n Gk ἀνά (ana) 'up', ὁδός (hodos) 'way'. Name given by Faraday in 1834 to the positive electrode towards which anions travel in electrolysis (see **cathode**).

ANODYNE n or adj Gk ἀν- (an-) 'not', ὀδύνη (odyne) 'pain'. A pain-relieving drug.

ANORTHITE n Gk ἀν- (an-) 'not', ὀρθός (orthos) 'right'. A plagioclase felspar mineral, $CaAl_2Si_2O_8$, distinguished from **orthoclase** (q.v.) by its cleavage.

ANTHOCYAN n Gk ἄνθος (anthos), 'flower', κύανος (cyanos) 'dark blue'. Name given by Marquart in 1835 to the blue colouring matters of flowers.

ANTHOCYANIDIN n See prec wd. A coloured substance (an aglycone) produced together with a sugar on hydrolysis of an **anthocyanin** (see foll wd).

ANTHOCYANIN n See prec wd. A member of one of the main classes of plant pigments found in flowers and fruits in the form of glycosides.

ANTHRACENE n Gk ἄνθραξ (anthrax) 'coal'. The aromatic hydrocarbon $C_{14}H_{10}$ found in coal tar by Dumas and Laurent in 1832 and named by Laurent in 1837.

ANTHRAQUINONE n From **anthra(cene)** and **quinone**. The quinone $C_{14}H_8O_2$ first obtained, by Laurent, by oxidation of anthracene with nitric acid.

ANTI- prefix Gk ἀντί (anti) 'against, opposite', (opp of **syn**). (See Gk prefixes, Appendix 4.)

ANTIBIOTIC n or adj Gk ἀντί (anti) 'against', βίος (bios) 'life'. A substance which destroys microorganisms. The term was first used in this sense by Waksman, the discoverer of streptomycin, in 1941.

ANTIMONY n Med L *antimonium*, probably fr Arabic *al*, 'the', *ithmid* 'stibium' (old name for antimony). Element at. no. 51, symbol **Sb** (from L *stibium*).

ANTIOXIDANT n Gk ἀντί (anti) 'against', and oxidant (see **oxidise**). A substance added to prevent or reduce oxidation.

APATITE n Gk ἀπάτη (apate) 'deceit'. The mineral $Ca_5(PO_4)_3F$, so called because it is easily mistaken for other minerals. Also the group of related minerals of empirical formula $M_5(XO_4)_3Y$ where M is a bivalent cation, e.g., Ca, Sr, X is a Group V element, e.g. P or As, and Y is a univalent anion, e.g. F, OH, Cl.

APOPHYLLITE n Gk ἀπό (apo) 'off, away from', φύλλον (phyllon) 'leaf'. A hydrated silicate mineral, containing calcium, potassium and fluoride ions, with sheet-like silicate structure, and so named because it exfoliates on heating.

APPARATUS n L *apparatus* 'equipment'. In chemistry, the equipment used for preparation, separation, analysis and other investigations.

APROTIC adj Gk ἀ- (a-) 'without', and **proton** (q.v.). Having no

25

strong tendency either to donate or to accept protons (a term used in description of solvents).

AQUA FORTIS L 'strong water'. Obsolete name for concentrated nitric acid, denoting its great solvent power.

AQUA REGIA L 'royal water'. Obsolete name for a mixture of nitric and hydrochloric acids, acquired on account of the ability of this mixture to dissolve gold (the 'king of metals').

AQUEOUS adj L *aqua*, 'water', *aquosus*, 'abounding in water'. Watery, consisting of water (as distinct from non-aqueous).

Ar Chemical symbol for **argon** (q.v.), element at. no. 18.

ARABINOSE n From *Gum arabic* and **-ose** (suffix denoting a sugar). A sugar first obtained by Scheibler in 1868 by boiling gum arabic with acid; an aldo-pentose $CHO(CHOH)_3 CH_2OH$.

ARACHIDIC adj Mod L *arachis*, fr Gk ἄραχος (arachos), name of a leguminous plant. Arachidic acid, the acid $C_{19}H_{39}COOH$, so named because of its occurrence in the Ground Nut, a member of the genus *Arachis*.

ARAGONITE n From *Aragon*, in Spain. A particular crystalline form of mineral calcium carbonate, distinct from the more common form **calcite**.

ARGENTUM n L *argentum*, 'silver', hence the chemical symbol **Ag** for **silver**, and the names of a number of argentiferous minerals.

ARGININE n Etymology uncertain. Name given in 1886 to the basic aminoacid $C_6H_{14}NO_2$ found in animal proteins and some vegetable tissues.

ARGON n Gk ἀργός (argos) 'idle', fr Gk ἀ- (a-) 'not', ἔργον (ergon), 'work, action'. The noble gas, element at. no. 18, discovered by Ramsay and Travers in 1894 and so named by them on account of its chemical unreactivity.

AROMATIC adj L *aroma*, fr Gk ἄρωμα (aroma) fragrance. A term applied originally to those substances, later shown to be benzene derivatives, found in certain aromatic substances, to distinguish them from the carbon compounds found in fats (see **aliphatic**). Term applied to all benzene derivatives and many other compounds containing ring systems with similar characteristics.

ARSENIC n L *arsenicum*, fr Gk ἀρσενικόν (arsenicon), by metathesis (and addition of *ov*) fr Pers *azzarñkh*, 'yellow orpiment' (*zar*, 'gold'). Chemical element at. no. 33, symbol **As**.

ARYL adj From **ar(omatic)** and **-yl**. Radical derived from an aromatic hydrocarbon, e.g. phenyl, C_6H_5, as distinct from **alkyl** (q.v.).

As Chemical symbol for **arsenic**, element at. no. 33.

ASBESTOS n L *asbestos*, fr Gk ἄσβεστος (asbestos), fr Gk ἀ- (a-), 'not', and σβεστός (sbestos), 'quenched'. Lit 'unquenchable, inextinguishable'. A group of silicate minerals of great importance on account of their fibrous nature and their incombustibility.

ASCORBIC ACID n Gk ἀ- (a-) 'without', and Med L *scorbutus*, 'scurvy'. Name given by Szent-Gyorgyi and Haworth in 1933 to the crystalline substance hexuronic acid, $C_6H_8O_6$, the anti-scorbutic vitamin commonly known as Vitamin C.

-ASE Name-ending signifying **enzyme** (q.v.).

ASPARAGINE n L, fr Gk ἀσπάραγος (asparagos) 'sprout'. An organic nitrogen compound, NH_2-$CH(CH_2CONH_2)COOH$, found in asparagus and first isolated by Robiquet in 1805.

ASPARTIC ACID n See prec wd. The aminoacid $HOOCCH_2CH(NH_2)$-$COOH$, closely related to **asparagine** and produced from it by hydrolysis.

ASPHALT n F *asphalte*, fr L *asphaltum*, fr Gk ἄσφαλτος (asphaltos) 'asphalt'. A black, plastic, combustible mineral substance, otherwise known as bitumen, and consisting of a mixture of hydrocarbons.

ASPIRIN n From Gk ἀ- (a-) 'without' and L *spiraea*, a plant. A hybrid name coined by Dreser in 1899 for synthetic acetylsalicylic acid, $C_6H_4(OCOCH_3)COOH$, to indicate its identity with, but different origin from, the natural substance found in *Spiraea ulmaria*.

ASTATINE n Gk ἀ- (a-) 'not', στατός (statos) 'standing', which give Gk ἄστατος (astatos) 'unstable'. Name given on account of its radio-active instability to element at. no. 85, the 'heaviest' member of the halogen group, symbol **At**.

ASYMMETRIC adj Gk ἀ- (a-) 'without', σύν (syn), 'with', μέτρον (metron), 'measure'. Not symmetric, having an unsymmetrical arrangement of its constituent parts.

ASYMMETRY n See prec wd. Lacking symmetry.

At Chemical symbol for **astatine**, element at. no. 85.

-ATE From L *-atus*, a suffix denoting function. Word-ending denoting 'salt of', as in acetate, sulphate, etc., salts of acetic, sulphuric acids, etc.

ATMOSPHERE n Gk ἀτμός (atmos) 'steam, vapour', σφαῖρα (sphaira) 'sphere'. The gaseous envelope of the earth or of any other body. (The term was actually coined by John Wilkins in 1638 in relation to the moon!)

ATOM n F *atome*, fr L *atomus*, fr Gk ἄτομος (atomos), fr Gk ἀ- (a-), 'without', τομός (tomos), fr τέμνειν (temnein) 'to cut'; i.e. 'indivisible'. The smallest complete particle of a chemical element.

ATOMIC adj See prec wd. Relating to atoms.

ATROPINE n From *Atropa*, 'deadly nightshade', fr Gk Ἄτροπος (Atropos), one of the Fates in Greek mythology, fr Gk ἀ- (a-), 'without', τρόπος (tropos), 'turn' i.e. 'inflexible'. An alkaloid, $C_{17}H_{23}NO_3$, found in deadly nightshade.

ATTO See SI prefixes. Prefix signifying 10^{-18} times the fundamental unit (cf. Danish *atten*, 'eighteen').

Au L *aurum*, 'gold'. Chemical symbol for **gold**, element at. no. 79.

AUFBAU adj Ger *Aufbau*, 'building up', fr Ger *auf*, 'up', *bauen*, 'to build'. The aufbau principle is concerned with the arrangement of the electrons in orbits round the nuclei of atoms.

AURIFEROUS adj L *aurum*, 'gold', *fero*, 'I bear'. Gold-bearing (in reference to rocks). See foll entry.

AURUM n L word for **gold**. Hence chemical symbol **Au** for gold, element at. no. 79.

AUTO- Before a vowel **aut-**, combining form denoting 'self, by oneself, independently'. Gk αὐτός (autos), 'self'.

AUTOCATALYSIS n Gk αὐτός (autos), 'self', κατά (cata), 'down', λύειν (lyein) 'to loosen'. **Catalysis** (q.v.) of a chemical reaction by the products of the reaction itself.

AUTOCLAVE n Hybrid word coined from Gk αὐτός (autos), 'self', with L *clavis*, 'key'. A self-locking pressure vessel used for sterilising or for carrying out chemical reactions under pressure.

AUTOXIDATION n From **auto-** (fr Gk, see prec wd), oxidation. Spontaneous oxidation of a substance by oxygen, part of which combines with the substance, whilst another part is converted into ozone or hydrogen peroxide or oxidises another substance.

AZA- From F *azote*, 'nitrogen', coined by Lavoisier in 1776 fr Gk α- (a-) 'without', ζωή (zoe), 'life', to denote the inability of nitrogen to support life. Prefix indicating the presence of a nitrogen atom in a chain or ring.

AZEOTROPE n From Gk prefix **a-**, 'without', ζεῖν (zein) 'to boil', and τροπή (trope), 'turning, change'. Lit. 'boiling without change'. A mixture of liquids which boils at a constant temperature.

AZIDE n See aza-. A compound containing the radical or ion N_3; a derivative of hydrazoic acid, HN_3.

AZINE n See aza-. A compound containing a six-membered ring of atoms of which one or more is nitrogen. (But note hydrazine, N_2H_4.)

AZO- See aza-. Chemical prefix indicating the presence of the functional group —N=N—.

AZOLE n See aza-. A compound containing five atoms in a ring, of which one or more is nitrogen.

AZURITE n From *azure*, 'clear blue' (fr L *azura*, fr Arabic *al-lazaward*, fr Persian *läjwärd*, 'lapis lazuli, ultramarine', fr name of the place *Lajward*, in Turkestan, from which ultramarine was obtained). The clear blue mineral copper carbonate, $2CuCO_3 . Cu(OH)_2$.

B

B Chemical symbol for **boron**, element at. no. 5.

Ba Chemical symbol for **barium**, element at. no. 56.

BACTERIA n Plural of **bacterium** (q.v.).

BACTERICIDE n Hybrid word fr Gk βακτήριον (bacterion), 'little rod, little stick', dim of βακτήρια (bacteria), 'rod, stick', and L-*acidum*, 'killing' fr *caedere*, 'to kill'.

BACTERIUM n Gk βακτήριον (bacterion) 'little rod, little stick'. A microscopic organism, so named on account of the appearance under the microscope.

BADDELEYITE n Named after its discoverer, *Baddeley*. Mineral zirconium dioxide, ZrO_2.

BAKELITE n Named after its inventor, *Baekeland* (1863–1944). Trade name of synthetic resin made from phenol and formaldehyde.

BAR n Gk βάρος (baros), 'weight'. A unit of pressure ($= 10^5$ N/m²).

BARBITURATE n Origin of name uncertain. An important class of medicinal substances containing the pyrimidine ring, used as sedatives and hypnotics; derivatives of barbituric acid (see foll wd).

BARBITURIC adj Origin of name uncertain. Barbituric acid is malonyl urea, the parent substance of the barbiturates.

BARIUM n Gk βαρύς (barys) 'heavy', so named by its discoverer Davy in 1808 on account of its occurrence in **barytes** (q.v.) or 'heavy spar', the dense mineral barium sulphate, $BaSO_4$. The alkaline-earth metal, element of at. no. 56; symbol **Ba**.

BARN n Name proposed humorously by Holloway and Baker in 1942 for a convenient unit of cross-sectional area of atomic nuclei in nuclear bombardment investigations. 10^{-24} cm² $= 10^{-28}$ m².

BAROMETER n From Gk βάρος (baros), 'weight' and **-meter** (q.v.). An instrument for measuring the pressure of the atmosphere.

BARYON n Gk βαρύς (barys), 'heavy', and **-on** (by analogy with electron, proton). In nuclear physics, a heavy particle.

BARYTA n From **barytes** (see foll wd). Barium oxide, BaO, or barium hydroxide, $Ba(OH)_2$.

BARYTES n Gk βαρύς (barys), 'heavy'. Mineral barium sulphate, $BaSO_4$, also known as heavy spar.

BASALT n L *basaltes*, 'a dark marble from Ethiopia'. A very dark igneous rock containing 50–60% silica.

BASE n L *basis*, fr Gk βάσις (basis) 'foundation, base', a term introduced into chemistry by Rouelle in 1754. Alkali, or other substance capable of neutralising acids.

BATHOCHROMIC adj Gk βάθος (bathos), 'depth, χρῶμα (chroma)', colour. Colour-deepening, i.e. moving an absorption band to a region of longer wavelength.

BAUXITE n F fr *Les Baux*, the name of the place in Provence, France, where it was found. A mineral, consisting of a mixture of hydrated alumina, the main source of aluminium.

Be Chemical symbol for **beryllium**, the metallic element of at. no. 4.

BENITOITE n After the *San Benito* River, California, U.S.A., where found. The mineral gemstone barium titanium silicate, $BaTiSi_3O_9$.

BENTONITE n After *Fort Benton*, Wyoming, U.S.A. A montmorillonitic clay mineral of importance in industry.

BENZ- (before a vowel) or **BENZO-** From **benzene**. Related to or derived from benzene (q.v.).

BENZAL adj See **benzene**. The radical $C_6H_5CH:$, e.g. as in benzalaniline, $C_6H_5CH:NC_6H_5$.

BENZALDEHYDE n From **benz(ene)** and **aldehyde**. The aldehyde C_6H_5CHO.

BENZEDRINE n From **benz(ene)** and **(ephe)drine** (q.v.). The physiologically active substance $C_6H_5CH(OH)CH(CH_3)NHCH_3$, used in medicine.

BENZENE n From **benzoic** (q.v.). Name coined by Mitscherlich in 1833 for the liquid obtained by distilling benzoic acid with lime. Originally

benzin and later benzol, the spelling benzene was introduced by Hofmann. The cyclic aromatic hydrocarbon C_6H_6, parent compound of a vast number of derivatives.

BENZIDINE n From **benz(ene)**. The compound 4,4'-diaminodiphenyl $H_2NC_6H_4C_6H_4NH_2$.

BENZIL n From **benz(ene)**, (see **benzoin**). The compound dibenzoyl, $C_6H_5COCOC_6H_5$.

BENZINE n See **benzene**. Confusingly, this name is not synonymous with benzene, but is used indiscriminately in reference to light petroleum consisting chiefly of aliphatic hydrocarbons.

BENZO- See **benz-**. The radical $C_6H_5C:$.

BENZOATE n See **benzoic (acid)**. Salt or ester of benzoic acid, e.g. sodium benzoate, C_6H_5COONa, ethylbenzoate, $C_6H_5COOC_2H_5$; the ion or radical C_6H_5COO.

BENZOIC adj From **benzoin** (q.v.). The acid, C_6H_5COOH, so named on account of its occurrence in gum-benzoin (see foll wd).

BENZOIN n F *benjoin*, fr Spanish *benjui*, which derives from *lo-benjui*, fr the Arabic *lubān jawī*, meaning frankincense of Java (Java being the former name of Sumatra). Benzoin, or gum-benzoin, is a balsamic resin obtained from the Sumatran tree Styrax benzoin. Benzoin is also the name of the substance $C_6H_5CH(OH)COC_6H_5$, found in gum-benzoin.

BENZOYL adj From **benzoin** and **-yl**. Name coined by Wöhler and

30

Liebig in 1832 for the radical C_6H_5CO which 'preserves its nature and composition unchanged in nearly all its associations with other bodies. This stability, this sequence in the phenomena, induced us to assume that this group is a compound element and hence to propose for it a special name, that of benzoyl'.

BENZYL adj From **benz-** and **-yl** (q.v.). The radical $C_6H_5CH_2$, as in benzyl chloride, $C_6H_5CH_2Cl$.

BERBERINE n L *berberis*, *barbaris*, 'barberry', fr Arabic *barbārīs* A physiologically active alkaloid $C_{20}H_{19}NO_5$, found in the root of the barberry.

BERKELIUM n After *Berkeley*, California, U.S.A. Name coined by Thompson, Ghiorso and Seaborg in 1949 after the city where the element was first made and identified. The radioactive transuranium element of at. no. 97, symbol **Bk**.

BERTHOLLIDE adj After *Berthollet* (1748–1822), who had put forward a theory of indefinite proportions. Name introduced in 1914 in reference to solid phases of variable composition.

BERYL n OF *beryl*, fr L *beryllus*, fr Gk βήρυλλος (beryllos), 'beryl'. A pale green precious stone, beryllium aluminium silicate, $Be_3Al_2Si_6O_{18}$.

BERYLLIUM n From **beryl** (see prec wd). The light metallic element, at. no. 4, symbol **Be**, found in beryl.

BETA adj Gk βῆτα (beta), second letter of Gk alphabet. Name applied in early studies of radioactivity (1899) to rays, more penetrating than X-rays, emitted by uranium and later shown to consist of fast-moving electrons.

BETAINE n L *beta*, 'beet'. Trimethylglycine, a colourless crystalline substance, $(CH_3)_3NCH_2COO$, found in beet by Scheibler in 1866.

Bi Chemical symbol for **bismuth**, element at. no. 83.

BI- prefix L *bi-* 'twice, double'. Prefix commonly used in chemistry to denote 'two'. (See Latin prefixes. Appendix 3.)

BIDENTATE adj L *bi-* 'twice, double', *dens*, gen *dentis*, 'tooth', or *dentatus* 'toothed'. Having two points of attachment, e.g. as in the case of the ligand ethylene diamine, $H_2NCH_2CH_2NH_2$, both nitrogen atoms of which can be directly linked to the central metal atom of a complex.

BIFILAR adj L *bi-*, 'twice, double', *filum*, 'thread'. Double-threaded, or double-wired.

BILIRUBIN n L *bilis*, 'bile', *ruber*, 'red'. An orange-red pigment, $C_{33}H_{36}N_4O_6$, the chief colouring matter of bile.

BINARY adj L *binarius*, 'consisting of two'. Dual, or compounded of two elements, e.g. a metal oxide, or a metal halide.

BIO- Gk βίος (bios), 'life'. Combining form meaning 'life', as in biology 'the study of life', and biochemistry (q.v.).

BIOCHEMISTRY n From **bio-** (q.v.), and **chemistry**. That part of chemistry concerned with the chemical processes of animal and plant life.

BIOTIN n Gk $\beta\iota\acute{o}\tau os$ fr $\beta\acute{\iota}os$ (bios) 'life'. The crystalline substance $C_{10}H_{16}N_2O_3S$ (Vitamin H) found in yeast and egg-yolk and so named by Kogl and Tonnis in 1936 because it is necessary for the healthy growth of yeast and other organisms.

BIPYRAMID n From **bi-** (q.v.), and **pyramid**. A solid figure consisting of two pyramids joined together.

BISMUTH n Ger *Bismuth* (now *Wismut*), spelt *bisemutum* by Agricola (1530). The metallic element of at. no. 83, symbol **Bi**.

BISPHENOID n L *bi-*, 'double', Gk $\sigma\phi\eta\nu o\epsilon\iota\delta\acute{\eta}s$ (sphenoeides), 'wedge-shaped'. Shaped like a double wedge.

BITTERN n OE *bitter* and Ger *bitter*, 'bitter'. The bitter-tasting solution left after the crystallisation of common salt from sea water.

BITUMEN n L *bitumen*, 'pitch'. A blackish thermoplastic mineral substance consisting of a mixture of hydrocarbons.

BIURET n Ger, fr L *bi-*, 'double', and **urea** (q.v.), so named by Wiedemann in 1847 because it is formed on heating urea by the condensation of two urea molecules, with elimination of ammonia. The substance $H_{22}NCONHCONH_2$.

Bk Chemical symbol for **berkelium**, the radioactive transuranium element at. no. 97.

BLENDE n Ger *Blende*, fr Ger *blenden*, 'to deceive', so named because although it resembles the lead ore galena it yields no lead. Mineral zinc sulphide, ZnS, alternatively known as sphalerite (which name also carries the connotation 'deceitful').

BORATE n See **boric** and **boron**. A salt or ester of boric acid, H_3BO_3.

BORAX n F *borax*, fr Arabic *baúraq*, fr Persian *būrah*, 'borax'. A soluble crystalline salt, sodium borate, $Na_2B_4O_7.1OH_2O$, found as a mineral and then sometimes also known as tincal.

BORAZINE n From **bor(on)** and **azo-**, denoting the presence of both boron and nitrogen in the molecule. The colourless liquid, $B_3N_3H_6$, of cyclic structure of alternating BH and NH.

BORAZOLE n See **bor(on)** and **azo-**. Alternative name for **borazine** (see prec wd).

BORAZON n From **bor(on)** and **azo**, denoting the presence of boron and nitrogen. A very dense form of boron nitride, BN, first made in 1957.

BORIC adj From **bor(on)** (q.v.). Relating to boron, as in the case of boric acid, H_3BO_3.

BORIDE n From **bor(on)** and **-ide** (q.v.). A binary compound of boron with another element, e.g. calcium boride; CaB_6.

BORNEOL n From *Borneo*; so named by Gerhardt in 1843 on account of its occurrence in the

Borneo camphor tree *Dryobalanops camphora*. A fragrant crystalline solid, $C_{10}H_{17}OH$, a secondary alcohol very closely related to camphor and belonging to the family of bicyclic monoterpenes.

BORON n From **bor(ax)** (q.v.) and **carb(on)**, name proposed by Davy in 1812 to denote the source of the element, borax, and the fact that the element is 'more analogous to carbon than to any other substance'. Gay-Lussac and Thenard had proposed 'bore' but Davy wrote 'Bore cannot with propriety be adopted in our language, though short and appropriate in the French nomenclature'. The non-metallic element, at. no. 5, symbol **B**.

Br Chemical symbol for **bromine**, the halogen element, at. no. 35.

BRASS n OE *braes*, possibly fr a Semitic source. The alloy of copper with zinc.

BRIMSTONE n ME *brinston*, fr *brinnen*, 'to burn', and *ston*, 'stone'. Old common name for sulphur.

BROMATE n From **bromine** (q.v.). Oxy-salt of bromine containing the ion BrO_3^-.

BROMIDE n From **bromine** (q.v.). Binary compound of bromine, e.g. sodium bromide, NaBr, or ester, e.g. ethyl bromide.

BROMINE n From F *brome*, fr Gk βρῶμος (bromos), 'stink', and suffix **-ine**, by analogy with chlorine and iodine; name given by Balard in 1826 to the newly-discovered element to denote its pungent odour. Actually, Balard originally suggested the name 'muride' but this proved unacceptable and with Balard's consent was changed to 'brome' by Vanquelin, Thenard and Gay-Lussac. The halogen element, at. no. 35, symbol **Br**.

BROMO- Combining form, derived from **bromine** (see prec wd), denoting 'bromine-containing', as in bromobenzene, C_6H_5Br.

BROMOFORM n The **brom(ine)** analogue of **chlor(oform)** (q.v.).

BRONZE n F fr Ital *bronzo*, ult fr Persian *biring*. Alloy of copper and tin.

BRUCINE n After *Brucea antidysenterica*, the tree in which it occurs. (The tree was named by Sir Joseph Banks after the explorer James Bruce (1730–1794).) A poisonous alkaloid, $C_{23}H_{26}N_2O_4$, closely related to strychnine.

BRUCITE n After *A. Bruce*, an American mineralogist. Mineral magnesium hydroxide, $Mg(OH)_2$.

BUFFER n or adj Eng *buffer*, a device for absorbing shocks. Term used to describe solutions capable of resisting change of pH on addition of acid or alkali.

BURETTE n F *burette*, 'a small carafe', dim of F *buire*, 'a kind of jug'. A graduated glass tube open at one end and fitted with a stop-cock at the other, for delivering measured quantities of liquid, originally devised by Gay-Lussac.

BUTADIENE n From **buta(ne)**, **di-** and **-ene** (q.v.). The diolefin derived from butane, $CH_2:CHCH:CH_2$.

33

BUTANE n Formed fr L *butyrum,* 'butter' (cf. butyric acid) with suffix -ane denoting a saturated hydrocarbon. Either of the paraffin hydrocarbons, C_4H_{10}, i.e. n-butane $CH_3CH_2CH_2CH_3$ or iso-butane $CH_3CH(CH_3)_2$.

BUTENE n From **buta(ne)** and **-ene**, indicating unsaturation. One of the olefinic hydrocarbons, C_4H_8.

BUTYL adj From **but(ane)** and **-yl** (q.v.). The radical C_4H_9, which may be n-butyl $CH_3(CH_2)_3$, i-butyl $(CH_3)_2CHCH_2$, sec-butyl, $CH_3CH_2CH(CH_3)$, or t-butyl, $(CH_3)_3C$.

BUTYRIC adj L *butyrum,* 'butter'. Name of the acid, C_3H_7COOH, of rancid odour first isolated by Chevreul (1786–1889) in 1817 and so named on account of its presence as glycerides in butter.

C

C Chemical symbol for **carbon,** element at. no. 6. Symbol for **coulomb,** unit of electric charge; symbol for heat capacity per mole.

c Symbol for velocity of light. Symbol for **centi-** (q.v.) (SI system).

c. or **ca.** L *circa,* 'about'. Approximately.

Ca Chemical symbol for **calcium,** element at. no. 20.

CACODYL n Gk κακώδης (cacodes), 'evil smelling' and suff **-yl**; name given by Berzelius on account of the foul odour of it and its derivatives. The radical $(CH_3)_2As$- and its dimer.

CADAVERINE n L *cadaver,* 'corpse'; so called because it is produced in the putrefaction of flesh. The diamine $H_2N(CH_2)_5NH_2$, one of the ptomaines.

CADINENE n From *cade oil,* fr F *cade,* fr Late L *catanus,* 'cedar juniper'. A laevorotatory sesquiterpene $C_{15}H_{24}$, so named on account of its occurrence in cade oil, i.e. juniper tar oil, obtained from the tree *Juniperus oxycedrus.*

CADMIUM n L *cadmia,* 'calamine', fr Gk καδμεία (cadmeia), 'cadmean earth', fr Κάδμος (Cadmos), 'Cadmus', the fabulous founder of Thebes who introduced the alphabet into Greece. Name given by Strohmeyer, discoverer of the element, in 1817, on account of its occurrence in *cadmia,* i.e. in the zinc mineral now called **calamine** (q.v.). The metallic element, at. no. 48, symbol **Cd.**

CAESIUM or **CESIUM** (U.S.) n L *caesius,* 'bluish grey', so called by its discoverers Bunsen and Kirchhoff in 1860 in allusion to the two blue lines in its emission spectrum. The rare alkali metal, at. no. 55, symbol **Cs.**

CAFFEINE n F *cafeine,* fr F *café,* 'coffee'. The vegetable alkaloid $C_8H_{10}N_4O_2$ so named by Runge in 1821 on account of its occurrence in coffee. The active principle of coffee and tea.

CALAMINE n F fr modern L *calamina* fr L *cadmia*, fr Gk καδμεία (cadmeia) 'cadmean earth', fr Κάδμος (Cadmos), 'Cadmus', the fabulous founder of Thebes; so named on account of its occurrence near Thebes. Mineral zinc carbonate, $ZnCO_3$.

CALCIFEROL n From *calciferous*, L *calx.* gen *calcis*, 'lime', and *fero*, 'I bring or bear' and **sterol** (see ergosterol); so named by Angus and coworkers in 1931 in allusion to its marked ability to promote the normal growth of bones. Vitamin D2, active against rickets.

CALCINATION n See foll wd. The process of strong heating or incineration, or burning to ashes, as in the conversion of chalk into lime.

CALCINE v F *calciner*, fr Med L *calcinare*, an alchemical term meaning to reduce to a calx, fr L *calx*, gen *calcis*, 'lime'. To roast, to heat strongly, to incinerate, to burn to ashes.

CALCITE n Ger *Calcit*, fr L *calx*, gen *calcis*, 'lime'. Name coined by mineralogist von Haidinger in 1849 for the common form of mineral calcium carbonate (cf. **aragonite**).

CALCIUM n L *calx*, gen *calcis*, 'lime', name coined for the metal newly isolated from lime by electrolysis. The alkaline-earth metal, at. no. 20, symbol **Ca**.

CALIFORNIUM n. From *California*, U.S.A., name coined by Thompson, Street, Ghiorse and Seaborg in 1940 in honour of the university and state where the research was done. The radioactive transuranium element discovered in 1940, at. no. 98, symbol **Cf**.

CALOMEL n F of doubtful origin, but may be from Gk καλός (calos) 'beautiful', μέλας (melas) 'black'. Mercurous chloride, Hg_2Cl_2.

CALORIE n F fr L *calor*, 'heat'; term originating in France in 1787. Unit of heat in the C.G.S. system, the quantity of heat required to raise the temperature of one gramme of water through one degree Celsius. Not an SI unit. 1 calorie (thermochemical)= 4·184 joule.

CALORIMETER n F, a hybrid word from L *calor*, 'heat', and Gk μέτρον (metron), 'measure', introduced by Lavoisier in 1789. Apparatus for measuring the heat changes e.g. in chemical reactions.

CALX (pl **calces**) n L *calx*, 'lime'. Archaic term for the ash or residue left after calcination of a mineral or a metal.

CAMPHANE n From **camph(or)** and **-ane** (q.v.). The saturated bicyclic terpene, $C_{10}H_{18}$, usually known as bornane.

CAMPHENE n From **camph(or)** and **-ene** (q.v.). A colourless crystalline terpene, $C_{10}H_{16}$, closely related to and convertible into **camphor** (q.v.).

CAMPHOR n F *camphre*, fr Med L *camphora*, fr Arabic *Kāfūr*, 'camphor'. A volatile aromatic crystalline substance, $C_{10}H_{16}O$, obtained from the camphor tree, *Laurus camphora*, by distillation.

CAMPHORIC adj From **camphor**. Related to camphor.

CAMPHORIC ACID n See prec wd. A dibasic acid $C_8H_{14}(COOH)_2$ obtained from **camphor** by oxidation.

CAPRIC adj L *caper*, gen *capri*, 'goat'. Capric acid, the old name, now superseded by decanoic acid, of $C_9H_{19}COOH$, a fatty acid of strong odour of goats, found as glycerides in cow's milk and goat's milk.

CAPROIC adj L *caper*, gen *capri*, 'goat'. Caproic acid, the old name, now superseded by hexanoic acid, of the fatty acid, $C_5H_{11}COOH$, of strong goat odour, found as glycerides in goat's milk.

CAPROLACTAM n From **capro(ic)** and **lactam** (q.v.). The compound cyclo-$[NH(CH_2)_5CO]$, the lactam of caproic acid, made industrially by isomerisation of cyclohexanone oxime for polymerisation to a linear polyamide of the nylon type.

CAPRYLIC adj L *caper*, gen *capri*, 'goat' and suff **-yl** and **-ic**. Caprylic acid, the old name, now superseded by octanoic acid, of the fatty acid, $C_7H_{15}COOH$, of unpleasant goat odour, found as glycerides in cow's milk and in goat's milk.

CARBAMATE n See foll wd. A salt or ester of carbamic acid, e.g. ethyl carbamate (urethane) $H_2NCOOC_2H_5$.

CARBAMIC adj From **carb(onic)** and **am(ide)** (q.v.), on account of carbamic acid being an amide of carbonic acid. Carbamic acid, NH_2COOH.

CARBAMIDE n From **carb(onic)** and **amide** (q.v.). The diamide of carbonic acid, $CO(NH_2)_2$, better known as **urea**.

CARBANION n From **carb(on)** and **anion** (q.v.). Negative ion, i.e. anion, derived in a formal sense from a hydrocarbon by the removal of a positively charged hydrogen atom (cf. carbonium).

CARBAZOLE n From **carb(on)** and **azo-** (q.v.). The compound $(C_6H_4)_2NH$, a constituent of coal tar.

CARBIDE n From **carb(on)** and **-ide** (q.v.). A binary compound of carbon, e.g. calcium carbide, CaC_2, or silicon carbide, SiC.

CARBINOL n From **carb(on)** and **-ol** (q.v.). Term coined by Kolbe in 1868 from **carbin** (his name for the methyl radical), for methanol and other alcohols derived from it by the substitution of the hydrogens of the methyl group by alkyl radicals; thus trimethylcarbinol, $(CH_3)_3COH$.

CARBOHYDRATE n From **carbo(n)** and **hydrate** in reference to the general formula $C_x(H_2O)_y$. A large group of naturally occurring organic substances, including sugars, starch and cellulose.

CARBOLIC adj L *carbo*, 'coal', *oleum* 'oil'. Carbolic acid, the name of the mixture of phenols, mostly C_6H_5OH, obtained by the fractional distillation of coal tar.

CARBON n F *carbone*, fr L *carbo*, gen *carbonis*, 'coal, charcoal', name coined by Lavoisier in 1789. The non-metallic element of at. no. 6, symbol C.

CARBONATE n From **carbon** and -**ate** (q.v.). Salt or ester of carbonic acid, e.g. calcium carbonate, $CaCO_3$.

CARBONIC adj From **carbon** and -**ic** (q.v.). Carbonic acid, H_2CO_3, the acid formed by combination of carbon dioxide with water.

CARBONIUM adj From **carb(on)** and -**onium** (by analogy with ammonium). Carbonium ions are positive ions, i.e. cations, derived in a formal sense from a hydrocarbon by the removal of a negatively charged hydrogen atom. Any organic ion containing positively charged carbon (cf. **carbanion**).

CARBONYL n or adj From **carbon** and suff -**yl** (q.v.). The radical CO, as in carbonyl chloride, $COCl_2$, or in compounds of the type nickel carbonyl, $Ni(CO)_4$.

CARBORUNDUM n From **carbo(n)** and **(co)rundum**. Trade name for the abrasive silicon carbide, SiC, first made by Acheson 1892, coined in reference to the fact that the material is harder than corundum but not as hard as diamond.

CARBOXYL n From **carb(on)**, **ox(ygen)** and -**yl** (q.v.). Name of the group, COOH, characteristic of many organic acids.

CARBOY Persian *qarāba*, 'a large bottle'. A large glass container for liquids.

CARCINOGEN n Gk καρκίνος (carcinos), 'crab, cancer', γεννᾶν (gennan), 'to bring forth'. A cancer-inducing substance.

CARDANOL n From *Anacardium occidentale*, the cashew-nut tree. A mixture of phenols, $C_6H_4(OH)C_{15}H_{25-31}$, derived from **anacardic** acid.

CARDOL n See prec wd. A mixture of resorcinol derivatives, $C_6H_3(OH)_2C_{15}H_{25-31}$, found in cashew-nut shell liquid.

CARNALLITE n After *von Carnall*. Mineral potassium magnesium chloride $KMgCl_3.6H_2O$, an important potassium mineral found in the Stassfurt salt deposits.

CAROTENE n Ger *Carotin*, fr L *carota*, 'carrot', and -**ene** denoting unsaturation (q.v.). The yellow unsaturated polyisoprenoid hydrocarbons $C_{40}H_{56}$ first found in carrots by Wackenroder in 1832.

CARVACROL n Ger *Carvacrol* (Schweizer, 1841), fr Mod L *carum carvi*, botanical name of caraway, and L *acris*, 'sharp', and -**ol** (q.v.). A phenol, $C_{10}H_{13}OH$, found in caraway, also in camphor.

CARVONE n From *Carum carvi*, 'caraway' and -one. A ketone, $C_{10}H_{14}O$, of the terpene family, found in caraway oil.

CARYOPHYLLENE n From *Caryophyllus*, 'clove' fr Gk καρυόφυλλον (caryophyllon), 'clove tree'. A bicyclic sesquiterpene, $C_{15}H_{24}$, found in oil of cloves.

CASEIN n L *caseus*, 'cheese'. The phosphoprotein of cheese and milk, first obtained pure by Berzelius in 1813.

CASSITERITE n Named by Beudant in 1832 fr Gk κασσίτερος

(cassiteros), 'tin'. The mineral tin-stone, SnO_2, found in Cornwall and elsewhere; the principal ore of tin.

CATA- prefix Gk κατά (cata), signifying 'down', or intensifying the meaning of the basic word.

CATALYSIS n From **cata-** (see prec entry) and λύσις (lysis), 'a loosening, setting free'. Name given by Berzelius in 1835 to the initiation or acceleration of a chemical reaction by a substance which itself undergoes no substantial change.

CATALYST n See prec wd. An agent which brings about catalysis.

CATAPHORESIS n From **cata-** (q.v.), and Gk φορεῖν (phorein), 'to carry or bring'. The movement of electrically charged colloid particles towards the cathode, in an electric field (see electrophoresis).

CATECHOL n From *Acacia catechu* (botanical name for an East Indian tree; fr Malay word *kachu*), and **-ol** (q.v.). A phenolic substance (1,2-dihydroxybenzene, discovered by Reinsch in 1839) obtained by dry distillation of *catechu*, a brown, brittle material obtained from the wood of certain Asiatic trees.

CATENANE n L *catena*, 'chain'. A compound in which two or more ring molecules are 'mechanically' linked together, as in a chain.

CATENATION n L *catenatus*, 'linked together' fr L *catena*, 'chain'. The joining together in a chain of a number of atoms of the same chemical element, a property outstanding in the case of carbon.

CATHETOMETER n Gk καθετήρ (catheter) 'something let down' and μέτρον (metron), 'measure'. A travelling microscope for measuring vertical distances.

CATHODE n Gk κάθοδος (cathodos) 'way down', fr κατά (cata) 'down', ὁδός (hodos) 'way'. Name introduced by Faraday in 1834 for the negative pole in electrolysis (cf. **anode**).

CATION n From **cat(a)** (q.v.), and **-ion** (q.v.). Name introduced by Faraday in 1834 for the positively charged ion in electrolysis (cf. **ion**, **anion**), since it moves towards the cathode.

CAUSTIC adj F *caustique*, fr L *causticus*, fr Gk καυστικός (causticos), 'burning, corrosive'. Burning, corrosive, destructive of animal tissue. Term applied especially to alkalis, esp. sodium hydroxide, 'caustic soda', and potassium hydroxide, 'caustic potash'.

Cd Chemical symbol for **cadmium** (q.v.) element at. no. 48.

Ce Chemical symbol for **cerium** (q.v.) element at. no. 58.

CEDROL n L *cedrus*, fr Gk κέδρος (cedros), 'cedar' and suff **-ol** denoting an alcohol. A sesquiterpene alcohol, $C_{15}H_{25}OH$, found in cedarwood oil.

CELESTINE n Ital., *celestino*, 'sky blue'. Mineral strontium sulphate, $SrSO_4$, so named on account of its occasional fine blue colour.

CELLOPHANE n Hybrid name coined from **cellulose** (q.v.) and Gk φαίνειν (phainein) 'to show'. Transparent glossy film made from cellulose.

CELLULOID n Hybrid name coined by Hyatt in 1871 from **cellulose** (q.v.) and Gk suff -oid, 'like, resembling'. A solid thermoplastic inflammable material made from cellulose nitrate and camphor.

CELLULOSE n F fr L *cellula*, 'small chamber'. A carbohydrate, $(C_6H_{10}O_5)_n$, which in nature forms the walls of plant cells, and which consists of parallel chains of glucose condensed together to form a fibrous structure.

CELSIUS n After *Anders Celsius* (1701–1744), inventor of a temperature scale. (The belief that Celsius invented the scale widely known as the Centigrade scale is in fact erroneous: the scale he proposed had zero for the boiling point of water and 100 for the melting point of ice. His scale was inverted by Christen in 1743. In Britain the scale is widely known as the Centigrade scale in spite of the official adoption of the name Celsius in 1948.)

CEMENT n and v OF *ciment*, fr L *caementum* 'stone chippings'. A powder which, mixed with water to give a plastic mass, sets hard. More generally, any substance used to make solid bodies stick together firmly.

CEMENTITE n From **cement**. Term introduced in 1888 in reference to the hardening of iron by means of 'cement carbon'. Hard brittle iron carbide, Fe_3C, responsible for the brittleness of cast iron, and present in steel.

CENTI- Combining form meaning 'one hundredth' (10^{-2}). L *centi-*, fr *centum*, 'a hundred'. Symbol (SI system), **c.**

CENTIGRADE adj From F, fr L *centum*, 'hundred', *gradus*, 'step'. Scale of temperature in which the difference between the freezing point and boiling point of water is divided into a hundred degrees (cf. **Celsius**).

CENTRIFUGE n or v From Mod L *centrifugus*, 'fleeing from the centre', coined by Newton fr L *centrum* 'centre', *fugere*, 'to flee'. Apparatus for separating suspended particles by spinning.

CERAMIC n or adj Gk κεραμικός (ceramicos) 'pottery, earthenware'. Pottery and other products made by the strong heating of clay or various other materials.

CERITE n From **cerium** (see foll wd). A rare mineral, cerium silicate, found in Sweden.

CERIUM n Named by Berzelius (1803) after the minor planet *Ceres* (itself named after Ceres, the Roman goddess of corn and harvests) which had been discovered just previously, in 1801. The lanthanide element, at. no. 58, symbol Ce, obtained from monazite sand.

CERMET n From **cer(amic)** and **met(al)**. A ductile, heat-resisting combination of ceramic substances and metal sintered together, used where resistance to high temperature and corrosion is required.

CEROTIC adj L *cera*, 'wax', fr Gk κηρός (ceros) 'wax'. Cerotic acid,

$C_{26}H_{53}OH$, an acid discovered in beeswax by Brodie in 1804.

CERUSSITE n F *ceruse*, fr L *cerussa* 'white lead', perhaps ultimately fr Gk κηρός (ceros) 'wax' in the sense 'waxen', i.e. 'white coloured'. Mineral lead carbonate $PbCO_3$.

CERYL adj From Gk κηρός (ceros) 'wax' and suff **-yl**. Ceryl alcohol, $C_{27}H_{55}OH$, is found in Chinese wax.

CESIUM n See **caesium**.

CETANE n See foll wd. Hexadecane, $C_{16}H_{34}$.

CETYL adj L *cetus*, 'whale'. The radical $C_{16}H_{33}$, as in cetyl alcohol (hexadecanol) $C_{16}H_{33}OH$, isolated from spermaceti (found in the head of the sperm whale) by Chevreul in 1817.

Cf Chemical symbol for **californium** (q.v.), element at. no. 98.

C.G.S. Abbreviation of centimetre-gramme-second, a system of units now superseded by the SI system (q.v.).

CHABAZITE (pron kabazite) n Gk χαβάζιε (chabazie) erroneously fr Gk χαλάζιε, voc of χαλάζιος (chalazios) 'resembling a hailstone'. A zeolite mineral, calcium aluminium silicate, $CaAl_2Si_4O_{12}.6H_2O$, so named on account of its appearance.

CHALCEDONY (pron kalsedony) n Etymology uncertain, but possibly so named because found near *Chalcedon*, a town in Asia Minor. A translucent variety of **quartz** (q.v.) used in jewellery.

CHALCOGEN (pron kalkogen) n Gk χαλκός (chalcos), 'copper, brass', γεννᾶν (gennan) 'to generate, produce'. Lit 'copper-forming', so called because found in copper ores (cf. halogen). General name for the Group VI elements O, S, Se, Te, Po. (Note absence of connexion with the word 'chalk'.)

CHALCOGENIDE n See prec wd. Compound of a **chalcogen** with another element (cf. halide).

CHALK n OE *cealc*, 'chalk', fr L *calx*, *calcem*, 'lime', fr Gk χάλιξ (chalix) 'stone'. Mineral calcium carbonate, $CaCO_3$, formed from the shells of small marine organisms.

CHALONE (pron kaylone) n Gk χαλῶν (chalon), 'slackening', and **(horm)one** (q.v.). A hormone which inhibits the action of certain organs or tissues.

CHALYBEATE (pron kalibyate) adj Mod L *chalybeatus*, fr Gk χάλυβος (chalybos), 'steel'. Containing iron in solution (in reference to spring water).

CHELATE (pron keelate) adj or n Gk χηλή (chele), 'claw', i.e. the pincer-like claws of lobsters, crabs and scorpions. Term introduced by Morgan and Drew in 1920 to denote complexes containing a closed ring of atoms, formed by the joining of the central metal atom to two different points in the ligand molecule. Examples of chelating agents are ethylenediamine, 1,10-phenanthroline, 8-hydroxyquinoline, dimethylglyoxime and ethylenediaminetetracetic acid (EDTA).

CHELATOMETRY (pron keelaitometry) n Gk χηλή (chele) 'claw' (see prec wd), $\mu\acute{\epsilon}\tau\rho\text{o}\nu$ (metron) 'measure'. Methods of volumetric analysis involving titrations of, or by, chelating agents.

CHELIDONINE (pron kellidonine) n From *Chelidonium majus*, botanical name of celandine. A colourless crystalline alkaloid, $C_{20}H_{19}NO_5$, found in the celandine.

CHEMICAL adj or n See **chemistry**. Related to chemistry, or a substance used or made in the operations of chemistry.

CHEMILUMINESCENCE n Ger *Chemiluminescenz* (1905); fr **chemi-(cal)** and luminescence (see **luminescent**). The production of light in the course of chemical reactions, otherwise than by incandescence.

CHEMISORPTION n From *chemi(cal)* and *sorption*. Adsorption (q.v.) of a gas or vapour onto a solid surface by chemical, as distinct from the weaker physical, forces.

CHEMISTRY n The word 'chemistry' appeared in English in the 17th century, and seems to have arisen out of the name of the predecessor of chemistry, alchemy. The word 'alchemy' is derived from the Arabic name of the art, *al-kimiya*, fr *al*, 'the', and Gk $\chi\eta\mu\epsilon\acute{\iota}\alpha$ (chemeia) or $\chi\eta\mu\acute{\iota}\alpha$ (chemia), the art of transmuting metals, e.g. as practised by the Egyptians. The exact significance of this is not agreed: it may be derived from Gk $X\eta\mu\acute{\iota}\alpha$ (Chemia) 'Blackland', the old name of Egypt, in reference to the black alluvial soil of the Nile valley as opposed to the light-coloured sand of the surrounding desert. Chemistry is the study of the composition and structure of all the different kinds of matter, and of the processes by which materials may be changed by the making and breaking of the bonds between atoms.

CHEMOTHERAPY n From *chemo-* 'chemical', and Gk $\theta\epsilon\rho\alpha\pi\epsilon\acute{\iota}\alpha$ (therapeia) 'healing', term introduced by Ehrlich (1907) to denote the treatment of disease by chemical substances which act on micro-organisms or malignant tissue.

CHINA CLAY n From *China*, the country where it was originally worked and used to make Chinaware (i.e. china). Also known as kaolin after the name of the mountain region *kauling* in China whence it was obtained. A fine, pure white clay produced by the decomposition of felspar and consisting of microcrystalline hydrated aluminium silicate $Al_2Si_2O_5(OH)_4$.

CHIRAL (pron kyral) adj Gk $\chi\epsilon\acute{\iota}\rho$ (cheir), 'hand', and suff *-al* fr L *-alis*; hybrid word coined by Lord Kelvin in 1894 to denote any geometrical figure which, like a hand, cannot be brought into coincidence with its mirror image.

CHLOR-, CHLORO- See **chlorine**. Chlorine-containing.

CHLORAL n Coined fr **chlorine** and **alcohol**, because obtained, by Liebig in 1831, by action of chlorine on alcohol. Trichloracetaldehyde, CCl_3CHO.

CHLORAMINE n From **chloro-** and **amine** (q.v.). Compound in which chlorine is directly linked to nitrogen, especially Chloramine-T (the T signifying 'toluene'), $CH_3C_6H_4SO_2NHCl$, used as an antiseptic.

CHLORAMPHENICOL n Name coined by abbreviation of the chemical name chlor-amide-phenyl-nitroglycol. An antibiotic $C_{11}H_{12}N_2O_5Cl_2$, extracted from *Streptomyces venezuelae* and made synthetically.

CHLORANIL n From **chlor(o)-** and **aniline** (q.v.). Compound originally made by action of potassium chlorate and hydrochloric acid on aniline. Tetrachlorobenzoquinone, $C_6Cl_4O_2$.

CHLORIDE n See **chlorine** and **-ide**. Salt or ester of hydrochloric acid, e.g. sodium chloride, NaCl, ethyl chloride, C_2H_5Cl.

CHLORINE n Gk $\chi\lambda\omega\rho\acute{o}s$ (chloros) 'pale green'. Name given by Davy in 1811 to the substance then newly recognised as an element and known under a number of other names from the time of its discovery by Scheele in 1774. The halogen element of at. no. 17, symbol **Cl**.

CHLORO- See **chlorine**. Chlorine-containing.

CHLOROFORM n F *chloroforme*, hybrid name coined by Dumas in 1834 from **chloro-** and F *formique*, 'formic', fr L *formica*, 'ant' (see **formic**) for the substance $CHCl_3$, because on hydrolysis with alkali it yields chloride and formate.

CHLOROMYCETIN n From **chloro-** and Gk $\mu\acute{u}\kappa\eta s$ (myces), gen $\mu\acute{u}\kappa\eta\tau os$ (mycetos), 'fungus'. Trade name of **chloramphenicol** (q.v.) the antibiotic, $C_{11}H_{12}N_2O_5Cl_2$, found in nature and manufactured synthetically.

CHLOROPHYLL n F, fr Gk $\chi\lambda\omega\rho\acute{o}s$ (chloros) 'pale green', $\phi\acute{u}\lambda\lambda o\nu$ (phyllon), 'leaf', coined by Pelletier and Caventou in 1819. The green colouring matter of plants, vital to photosynthesis; actually a mixture of $C_{55}H_{72}N_4O_5Mg$ and $C_{55}H_{70}N_4O_6$-Mg.

CHLOROPICRIN n From **chloro-** and **picric acid** (q.v.) because made from chlorine and picric acid. The pungent, toxic substance CCl_3NO_2.

CHOLESTEROL n Gk $\chi o\lambda\acute{\eta}$ (chole), 'gall bile', $\sigma\tau\epsilon\rho\epsilon\acute{o}s$ (stereos) 'solid', and **-ol**, denoting an alcohol. So called because originally found in gall-stones. The solid unsaturated alcohol, $C_{27}H_{45}OH$, present in all animal tissues, especially in the brain, spinal cord and gall-stones.

CHOLINE n Gk $\chi o\lambda\acute{\eta}$ (chole), 'bile'. The crystalline base $HO(CH_2)_2$-$N(CH_3)_3OH$, found in bile by Strecker in 1862. A member of the Vitamin B complex.

CHROMATOGRAPHY n Gk $\chi\rho\hat{\omega}\mu a$ (chroma), 'colour', $\gamma\rho\acute{a}\phi\epsilon\iota\nu$ (graphein), 'to write'. A method of chemical analysis so named because in its early applications mixtures of coloured substances in solution were separated during passage through a column of adsorbent, giving rise to bands of colour. The technique is widely used for separation of mixtures of volatile substances carried in a stream of inert gas as well as for mixtures of soluble substances dissolved in a liquid.

CHROMIUM n Gk $\chi\rho\hat{\omega}\mu a$ (chroma) 'colour'. Name coined by Fourcroy and Haüy in allusion to the colours of its compounds. The metallic element, discovered by Vanquelin in 1797, of at. no. 24, symbol **Cr**.

CHROMOPHORE n From Gk χρῶμα (chroma), 'colour' and -φόρος (phoros), 'carrying'. A chemical grouping in an aromatic compound which confers colour on the compound.

CHROMYL adj From **chromium** (q.v.). The radical CrO₂, as in chromyl chloride, CrO_2Cl_2 (cf. sulphuryl).

CHRYSOTILE n Gk χρυσός, (chrysos) 'gold', so named originally because some specimens of the mineral have the appearance of fine golden hair, the colour being due to the presence of minute amounts of iron impurity. The chief asbestos mineral used in industry, a magnesium silicate $Mg_3Si_2O_5(OH)_4$.

CINCHONIDINE n From *Cinchona*, botanical name of a tree the bark of which yields a cure for fevers. Originally growing in Peru, the tree is now cultivated in India and Java for the sake of its bark. An alkaloid, $C_{19}H_{22}N_2O$, related to quinine and found with it in the bark of the Cinchona tree.

CINCHONINE n See prec wd. An alkaloid occurring with cinchonidine and isomeric with it.

CINEOL n L *oleum cinae*, 'oil of wormseed' fr *Artemisia cina*, botanical name of wormseed. A liquid terpene $C_{10}H_{18}O$ with a characteristic camphor-like odour; the chief constituent of wormseed oil.

CINNABAR n L *cinnabaris*, fr Gk κιννάβαρι (cinnabari), 'cinnibar', probably fr Persian *'zanzifrah'*. Red mineral mercury sulphide, HgS, the chief ore of mercury.

CINNAMIC adj L *cinnamomum*, fr Gk κίνναμων (cinnamon), 'cinnamon', fr Hebrew *qinnāmōn*. Cinnamic aldehyde, $C_6H_5CH:CHCHO$, is the chief constituent of oil of cinnamon. Cinnamic acid $C_6H_5CH:CH COOH$ is readily formed from the aldehyde by oxidation by air.

CIS L *cis* 'on this side of' (cf. *trans*, 'on the other side of'). The terms cis- and **trans**-, signifying 'on the same side' and 'on opposite sides', respectively, were introduced into chemistry by Baeyer in 1890 in the course of his work on hexahydrophthalic acid, of which he prepared the cis- and trans- forms. Cis-trans isomerism occurs both in organic chemistry (e.g. maleic and fumaric acids) and in inorganic chemistry (e.g. in $Pt(NH_3)_2Cl_2$).

CITRIC adj L *citrus*, the name now used for lemon, orange and similar trees. (In ancient times it had a different significance, and the name derives from Gk κέδρος (cedros) 'cedar'.) Citric acid, $HOOCCH_2-CH(OH)(COOH)CH_2COOH$, is so named on account of its presence in the juice of citrous fruits.

CITRONELLAL n From *Citronella* oil, and -al (q.v.) denoting an aldehyde. A naturally occurring terpene aldehyde, $C_9H_{17}CHO$, with a pleasant lemon odour, found in citronella oil.

Cl Chemical symbol for **chlorine** (q.v.) element at. no. 17.

CLATHRATE n or adj L *clathri*, 'trellis, or grille'. Term introduced by H. M. Powell in 1948 to designate a new type of compound in which the components are bound together not by chemical bonds but by the trapping or imprisonment of molecules of one sort in a framework of the other, e.g. argon in a framework of quinol molecules.

Cm Chemical symbol for **curium** (q.v.), element at. no. 96.

Co Chemical symbol for **cobalt** (q.v.), element at. no. 27.

COBALT n Ger *Kobalt*, fr Ger *Kobold*, 'an evil sprite', so called because its ore could not be made to yield any useful metal (copper). *Kobold* is related to the Gk κόβαλος (cobalos), 'rogue, evil spirit' and the English *goblin* (cf. the similar derivation of **nickel**). The metal, transition element of at. no. 27, symbol **Co**.

COCAINE n From *coca*, the plant from which it is obtained. An alkaloid, $C_{17}H_{21}NO_4$, used as a local anaesthetic.

CODEINE n Gk κώδεια (codeia), 'poppy-head', in reference to its origin in opium. The alkaloid $C_{18}H_{21}NO_3$, used as an analgesic and hypnotic.

COESITE n After *L. Coes, jr.*, who first described it in 1953. A high density form of silica, SiO_2, discovered as a mineral.

COLCHICINE n From *Colchicum autumnale*, the botanical name of Meadow Saffron, or Autumn crocus, from which colchicine is obtained.

An alkaloid, $C_{22}H_{25}NO_6$, used in medicine and agriculture.

COLLAGEN n F *collagène*, fr Gk κόλλα (colla), 'glue', and γεννᾶν (gennan), 'to generate, produce'. A protein abundant in animal connective tissue and which yields gelatin on boiling with water.

COLLIDINE n Gk κόλλα (colla), 'glue', εἶδος (eidos), 'form', so named because originally found in the oil distilled from bones. A basic substance, $C_8H_{11}N$, a derivative of pyridine.

COLLIGATIVE adj L *colligare*, 'to bind together', fr col- (see prefixes) and *ligare* 'to bind'. Colligative properties are those diverse properties of dilute solutions (such as elevation of boiling point or osmotic pressure), which can be embraced in the generalisation that their magnitude is determined by the number of particles per unit volume and independently of their nature.

COLLODION n Gk κολλώδης (collodes), 'glue-like'. A gummy solution of cellulose nitrate in a mixture of ether and alcohol.

COLLOID n Gk κόλλα (colla), 'glue', -οειδης (-oeides), 'like' 'of the nature of glue'. A term introduced by Graham in 1861 to distinguish those substances which dialysed slowly, e.g. glue, from those which dialysed rapidly, e.g. simple salts, (which he classed as crystalloids).

COMANIC ACID n See **comenic** acid. The heterocyclic compound $C_5H_4O_4$, which can be regarded or derived fr comenic acid by removal of an OH group.

COMENIC ACID n Coined by Berzeluis in 1852 by metathesis fr **meconic acid**, (q.v.), whence it can be obtained by boiling with HCl. A six-membered heterocyclic compound, $C_6H_4O_5$.

COMPOUND n Mid E *compounen*, 'to mix', fr Old F *componre* or *compondre*, fr L *componere*, 'to put together'. A combination of elements, a substance made up of elements combined together.

CONIINE n From *Conium maculatum*, botanical name of hemlock, from which coniine is obtained. An alkaloid, 2-n-propylpiperidine, $C_8H_{17}N$, the active principle of hemlock.

CONJUGATE adj From L *conjugatus*, pp of *conjugare*, 'to join together', from L prefix con- (q.v.) and *jugare*, 'to join, to yoke'. According to the Lowry–Brønsted theory the relation between an acid and a base is expressed as A(acid) $\rightleftharpoons$ H$^+$(proton)+B(base); B is termed the conjugate base of the acid A.

CONJUGATED adj See previous entry. Term applied in organic chemistry to systems of alternating single and double bonds, which have properties distinct from those of either single or double bonds separately.

COPPER n L *cuprum*, 'copper', fr *aes Cyprium*, 'Cyprus brass', so called because Cyprus was an important source of the metal. Transition element, at. no. 29, symbol **Cu**.

CORTISONE n Name formed by abbreviation of its systematic chemical name, 17-dehydrocorticosterone.

A steriod hormone found in the cortex of the adrenal gland.

CORUNDUM n From the Tamil wd *kurundam*, 'ruby'. A crystalline form of alumina, Al_2O_3, which occurs in nature and which, because of its great hardness, is also manufactured for use as an abrasive.

COULOMB n After *C. A. Coulomb* (1736–1806), the French scientist. The unit of electric charge; the quantity of electricity transported in one second by a current of one ampere.

COUMARIN n F *coumarine*, fr *cumarú*, the native name in the Tupi language of Guiana for the tonka bean; so called on account of its occurrence in the seed of the tonka bean. A colourless crystalline substance, $C_9H_6O_2$, with a pleasant odour of vanilla.

COUMARONE n Ger *Cumaron*, named after **coumarin** (see prec wd). A compound, C_8H_6O, benzofuran, the oxygen analogue of indole, found in coal tar.

Cr Chemical symbol for **chromium** (q.v.), element at. no. 24.

CREATINE n Gk κρέας (creas), 'flesh'. An organic base, $C_4H_9N_3O_2$, isolated from extract of meat by Chevreul in 1835.

CREATININE n Gk κρέας (creas) 'flesh'. An organic base, $C_4H_7N_3O$, closely related to **creatine** (q.v.) and found with it in the juice of flesh.

CREOSOTE n Gk κρέας (creas), 'flesh', σώζω (sozo) 'preserve, save'. Name coined by Reichenbach in 1833 to express his belief that it was 'creosote' present in wood tar and wood smoke that preserved smoked meat from decay. Creosotes from wood tar and from coal tar respectively are complex mixtures of different compositions, both containing cresols. (See next wd.)

CRESOL n From **creosote** and suff -ol (q.v.) denoting a phenol. An aromatic liquid phenol obtained from coal tar and present in wood tar (see prec wd). The three isomeric cresols or hydroxytoluenes, $CH_3C_6H_4OH$.

CRISTOBALITE n Ger *Cristobalite*, after *Cerro San Cristobal*, Mexico, where first found. One of the main crystalline forms of silica SiO_2.

CROCIDOLITE n Gk κροκύς (crocys), gen κροκύδος (crocydos) 'a nap of woollen cloth', λίθος (lithos) 'stone', so called on account of its appearance. A fibrous mineral of the asbestos group.

CROCOISITE or **CROCOITE** n Gk κροκόεις (crocoeis), 'saffron-coloured'. Mineral lead chromate, so called on account of its colour.

CROCONIC adj Gk κρόκος (crocos), 'saffron'. Croconic acid, $C_5H_2O_5$, is so called in allusion to the saffron colour of the acid and its salts.

CROTONIC adj fr *croton oil*, an oil extracted from the seeds of *croton tiglium*, an Indian plant. Crotonic acid, an unsaturated acid, $CH_3CH:CHCOOH$, occurring in cis- and trans- forms, was first found in croton seeds.

CRUCIBLE n Med L *crucibulum*, 'melting pot; fr L *crux*, gen *crucis*, 'cross'. A melting pot, so named on account of its original shape, which suggested a cross. A heat resisting vessel for use at high temperatures.

CRYOGENICS n From Gk κρύος (cryos), 'frost' and γεννᾶν (gennan), 'to generate, produce'. The production and study of phenomena which occur at very low temperatures.

CRYOLITE n Gk κρύος (cryos), 'frost', λίθος (lithos), 'stone', so named on account of its translucent appearance. Mineral sodium aluminium fluoride, $Na_3Al F_6$.

CRYOSCOPY n Gk κρύος (cryos), 'ice, frost', σκοπεῖν (scopein), 'to examine'. The determination of molecular weights by observing the depression of freezing point of a solvent by a known concentration of solute (cf. **ebullioscopy**).

CRYSTAL n Gk κρύος (cryos), 'frost', and κρύσταλλος (crystallos), 'clear ice'. Originally, an ice-clear mineral, e.g. rock crystal. A solid body of definite geometrical form resulting from the regular arrangement of its ultimate particles.

CRYSTALLINE adj See prec wd. Having regular arrangement of its ultimate particles (cf. **amorphous**).

Cs Chemical symbol for **caesium** (q.v.), element at. no. 55.

Cu Chemical symbol for **copper** (L *cuprum*) (q.v.), element at. no. 29.

CUMENE n L *cuminum*, Gk κύμινον (cuminon), 'cumin', name of an umbelliferous plant, Roman caraway. Isopropyl benzene, $C_6H_5CH-(CH_3)_2$, a liquid hydrocarbon found in cumin oil.

CUPEL n F *coupelle*, fr L *cupella* 'a little tub'. A bone-ash dish used in the assay of noble metals.

CUPFERRON n Ger, fr L *cuprum* and *ferrum*. Name coined by Baudisch in 1909 for the ammonium salt of phenylnitrosohydroxylamine to indicate its usefulness as an analytical reagent for copper and iron.

CUPRITE n L *cuprum* 'copper' (q.v.) and suff -ite. Mineral copper oxide, Cu_2O.

CURARINE n From '*curare*', fr the Tupi word *urari* (which, according to Klein, means 'he to whom it comes, falls'). The alkaloid, $C_{19}H_{26}-N_2O$, obtained from curare, the very poisonous black resinous substance used by the Indians of South America to tip their arrows.

CURIE n After *Marie Curie* (1867–1935). A measure of radioactivity; the quantity of radioactive substance which decays at the rate of $3·7 \times 10^{10}$ disintegrations per second.

CURIUM n After *Pierre* and *Marie Curie*. Name coined by its discoverers, Ghiorso, James, Morgan and Seaborg, in 1945 for the transuranium element, at. no. 96, symbol **Cm**.

CYANAMIDE n From *cyan-* (see **cyanogen**) and **amide** (q.v.). The compound $CNNH_2$ and its derivatives, e.g. calcium cyanamide $CaNCN$.

CYANATE n From **cyan-** (see cyanogen) and **-ate** (q.v.). A salt of cyanic acid.

CYANIDE n See **cyanogen**. Compound containing the radical or ion CN; a derivative of hydrogen cyanide, HCN.

CYANIN n Gk κύανος (cyanos) 'dark blue'. The colouring matter of the blue cornflower and other flowers.

CYANOCOBALAMINE n By abbreviation from its chemical name. Vitamin B_{12}, $C_{63}H_{90}N_{14}PCo$.

CYANOGEN n F *cyanogène*, fr Gk κύανος (cyanos), 'dark blue', and *-gen*, 'producing', fr Gk -γενής (genes), 'produced by'. The poisonous gas, $(CN)_2$, so named on account of the part played by cyanide in the pigment Prussian blue.

CYCLIC adj Gk κύκλος (cyclos), 'circle'. Consisting of a ring of atoms in the molecule, e.g. cyclohexane, $(CH_2)_6$, benzene C_6H_6, pyridine, C_5H_5N, etc.

CYCLO- Combining form of **cyclic** (see prec wd), e.g. cyclohexane $(CH_2)_6$ as distinct from the open chain of n-hexane, C_6H_{14}.

CYCLOTRON n Gk κύκλος (cyclos), 'circle', and *-tron* (by analogy with electron, etc.). Apparatus for accelerating charged particles by causing them to travel in a spiral path between two hollow semicircular electrodes, the particles thereby receiving an increase of energy at each half-revolution.

CYMENE n From *Cuminum cyminum*, botantical name of the plant from which oil of cummin is obtained. The liquid hydrocarbon, 4-isopropyltoluene, $CH_3C_6H_4CH(CH_3)_2$, found in cummin oil.

CYSTEINE n See **cystine**. A sulphur-containing amino-acid, $HSCH_2CH(NH_2)COOH$, found in most proteins and closely related to cystine.

CYSTINE n Gk κύστις (cystis) 'bladder', so called because found in a kind of urinary calculus (a bladder stone). A sulphur-containing amino-acid, $[SCH_2CH(NH_2)COOH]_2$, closely related to cysteine.

CYTOCHEMISTRY n Gk κύτος (cytos), 'a hollow vessel, a cell' and **chemistry** (q.v.). That part of chemistry concerned with living cells.

CYTOSINE n Gk κύτος (cytos), 'a hollow vessel, a cell'. A pyrimidine base, $C_4H_5N_3O$, found in the nucleotides of the nucleic acids of living cells.

D

D Chemical symbol for **deuterium** (q.v.), i.e. for hydrogen of mass number 2. Symbol for diffusion coefficient.

d Symbol for diameter, and for relative density (IUPAC); for differ-

ential; and denoting angular quantum number $= 2$ (indirectly arising from the 'diffuse' series in the atomic spectra of the alkali metals). Symbol for **deci-**(10^{-1}) in the SI system.

d- Abbreviation of **dextro-** (q.v.) now superseded by $(+)$.

da Symbol for **deca-** $(10 \times)$ in the SI system.

DALTON n After *John Dalton* (1766–1844). The atomic mass unit, i.e. the unit of the scale in which the mass of the carbon isotope, $^{12}_{6}C$, is taken to be 12 exactly; thus one dalton is equivalent to $1 \cdot 660149 \times 10^{-27}$ kg.

DDT Abbreviation of dichlorodiphenyltrichlorethane, an insecticide.

DE- Prefix fr F and fr L, meaning (i) down; (ii) away from; (iii) completely; (iv) opposite of.

DEBYE UNIT After *P. J. W. Debye* (1884–1966). A unit of molecular dipole moment equal to 10^{-18} electrostatic units or $3 \cdot 336 \times 10^{-30}$ coulomb metre.

DECA- Prefix denoting $10 \times$ in the SI system of units. Gk δέκα (deca), 'ten'.

DECALIN n From *decahydronaphthalene*. A saturated bicyclic hydrocarbon, $C_{10}H_{18}$, made by catalytic hydrogenation of naphthalene.

DECANT v F *décanter*, fr Med L *decanthare*, 'to pour from the lip of a jug'. To pour off a liquid gently so as not to disturb the sediment.

DECI- Prefix denoting 10^{-1} in the SI system of units. F *deci*, arbitrarily formed fr L *decimus*, 'the tenth'.

DEGENERACY n L *degenare*, lit. 'to depart from one's race', from **de-** (q.v.) and *genus*, gen. *generis*, 'birth race'. Signifies that a given energy level in a system of molecules is occupied by two or more molecules, or in an atom that the given energy level corresponds to two or more sets of quantum numbers.

DEHYDRATE v From **de-** 'to deprive of' and **hydrate** (q.v.). To deprive of water, to eliminate or remove water, including chemically combined water.

DELIQUESCE v L *deliquescere*, 'to dissolve, to melt away', fr **de-**, 'down', *liquescere*, 'to become liquid'. To become liquid by absorption of moisture from the air.

DELPHINIDIN CHLORIDE n From *Delphinium staphisagria*, 'larkspur'. A pigment, $C_{15}H_{11}O_7Cl$, present in the wild purple larkspur and the blue-black pansy.

DENDRITIC adj Gk δένδρον (dendron), 'tree'. A growth of mineral or crystals showing many branches and resembling a tree.

DEOXY- Combining form indicating removal of oxygen (see also **desoxy-**).

DEOXYRIBONUCLEIC ACID n From 2-**deoxy** D-**ribose**, the sugar involved, and **nucleic acid** (q.v.). The substance in chromosomes and viruses consisting of long molecules composed of two interwound poly-nucleotide chains, each nucleotide consisting of the sugar combined with one of four bases, adenine, cytosine, guanine and thymine. The molecules carry the genetic code. Abbreviation **DNA**.

DEPSIDE n Gk δέψειν (depsein), 'to tan'. A condensation product of substituted hydroxybenzoic acids, so named by Fischer because of the general resemblance to tannins. Many depsides are found in lichens.

DESICCATOR n L *desiccatus*, 'dried up completely', fr **de-**, 'completely', *siccare*, 'to make dry'. A vessel for effecting 'complete' dehydration.

DESORPTION n **de-** 'opposite of', **sorption** (q.v.). Reverse process of absorption and adsorption; yielding up of sorbed matter.

DESOXY- Prefix indicating 'loss of oxygen' (see **deoxy-**).

DETERGENT n L *detergere*, 'to cleanse, to wipe away'. A cleansing agent.

DEUTERIUM n Gk δεύτερος (deuteros), 'second'. The second form of hydrogen (cf. protium and tritium). The isotope of hydrogen of mass number 2, i.e. ^2_1H, symbol **D**.

DEUTERON n See **deuterium**. The nucleus of the deuterium atom (by analogy with proton).

DEXTRIN n From *dextro-rotatory* (see next entry). The gummy dextrorotatory mixture of hydrolysis products of starch.

DEXTRO- L *dexter*, 'on the right-hand side'; abbreviation of 'dextro-rotatory', a term introduced by Fischer signifying the rotating of the plane of vibration of polarised light to the right (when viewed against the direction of the light). Opposite of laevo-rotatory. Abbreviation before a chemical name, **d-**, now superseded by (+).

DEXTROSE n From **dextro-** (see prec entry) and **-ose**, signifying 'sugar'. The dextro-rotatory mono-saccharide, $C_6H_{12}O_6$, commonly known as glucose, or grape-sugar.

DI- Gk δι- (di-), 'twice'. See Gk numerals, Appendix 2.

DIA- Gk διά (dia) 'through, asunder'. See Gk prefixes, Appendix 4.

DIALYSIS n Gk διά (dia), 'through', λύσις (lysis) 'loosening, setting free'. Term introduced by Graham in 1861 in reference to the separation of colloids and 'crystalloids' by means of their different rates of diffusion through a semipermeable membrane.

DIAMOND n F *diamant*, fr L *diamantem*, acc of *diamas*, blend of Gk ἀδάμας (adamas) 'adamant', (i.e. invincible), 'diamond', and διαφανής (diaphanes), 'transparent'. The allotrope of carbon esteemed as a gem and valued as an abrasive on account of its extreme hardness.

DIAPHRAGM n L *diaphragma*, fr Gk διά (dia), 'through', φράγμα (phragma) 'fence, screen'. A partition or dividing membrane.

DIASPORE n Gk διασπορά (diaspora), 'scattering about', fr διά (dia), 'asunder', σπείρειν (speirein), 'to scatter'. Mineral aluminium oxy-hydroxide, AlOOH, so named on account of its marked decrepitation on heating.

DIASTASE n Gk διάστασις (diastasis), 'making a breach', fr διά (dia), 'asunder', στάσις (stasis) 'placing'. Name coined by Payen and Persoz in 1833 for the enzyme in malt which converts starch into sugar during brewing in the belief that diastase pierced a skin round the starch granules and allowed the liquid contents to escape.

DIASTEREOMER or **DIASTEREOISOMER** n Gk διά (dia), 'through', στερεός (stereos), 'solid', ἴσος (isos) 'equal', μέρος (meros), 'part'. Two molecules are said to be diastereomers when, being isomeric, their configurations are the same at one or more asymmetric centres and different at other asymmetric centres. Two isomeric salts are said to be diastereomers when they consist of the combination of an optically active base with an optically active acid, and when the configurations of one of the ions, e.g. the cation, are the same in the two salts, whilst the configurations of the other ion, e.g. the anion, are different. Such pairs of salts may be separated by fractional crystallisation.

DIATOM n Gk διά (dia), 'through, asunder', τέμνειν (temnein), 'to cut', giving διάτομος (diatomos), 'cut through'. One of the microscopic unicellular algae, so named on account of the 'cut-through' appearance when viewed through the microscope.

50

DIATOMACEOUS adj From prec wd. Consisting of diatoms, e.g. kieselguhr, diatomaceous earth.

DIATOMIC n Gk δι- (di-), 'twice', and **atomic** (q.v.). Consisting of two atoms, e.g. the molecules H_2, N_2, O_2, Cl_2.

DIATOMITE n See diatomaceous. Diatomaceous earth, a siliceous deposit consisting of the fossils of diatoms.

DIAZO- Gk δι- (di-), 'twice', and **azo-** (q.v.). Combining form denoting the presence of two adjacent nitrogen atoms in the molecule, e.g. diazomethane, CH_2N_2.

DIAZONIUM adj Gk δι- (di-), 'twice', **azo-** (q.v.) and **-onium**, by analogy with ammonium. Name of the positive ion RN_2^+ (R = aromatic radical), present in diazonium salts.

DIBORANE n Gk δι- (di-), 'twice', **bor(on)** (q.v.) and **-ane** (q.v.) by analogy with ethane. The boron hydride B_2H_6.

DIDYMIUM n Gk δίδυμος (didymos), 'twin'. Name given by Mosander in 1841 to a supposed new element on account of its invariable association with lanthanum. Didymium was later (in 1885) shown to be a mixture of two very similar elements, **neodymium** (q.v.) and **praseodymium** (q.v.).

DIENE n Gk δι- (di) 'twice', and **-ene** denoting unsaturation (q.v.). A compound containing two double bonds in the molecule, e.g. butadiene, $CH_2:CHCH:CH_2$.

DIENOPHILE n From **diene** (see prec wd) and Gk φίλος (philos), 'loving'. A reagent which combines with dienes, e.g. maleic anhydride.

DIMER n Gk δι- (di-), 'twice', μέρος (meros), 'part'. A molecule composed of two identical simple molecules joined together (cf. monomer, oligomer, polymer).

DIMORPHISM n Gk δι- (di-), 'twice', μορφή (morphe), 'shape, form'. The occurrence of a substance in two different crystalline forms, e.g. calcium carbonate in the forms of calcite and aragonite.

DIOXAN n From Gk δι- (di-), 'twice', **'ox(ygen)'** and **-an** (q.v.). Saturated compound composed of a ring of four carbon atoms and two oxygen atoms, e.g. the cyclic diether, 1,4-dioxan, used as a solvent.

DISACCHARIDE n Gk δι- (di-), 'twice', σάκχαρον (saccharon), 'sugar'. A sugar such as sucrose, cane sugar, $C_{12}H_{22}O_{11}$, formed from two monosaccharide units linked together.

DNA Abbreviation of **deoxyribonucleic acid** (q.v.).

DODECA- Gk δώδεκα (dodeca), 'twelve' (see Gk prefixes, Appendix 4).

DOLOMITE n After *de Dolomieu* (1750–1801) a geologist. Mineral calcium magnesium carbonate (Ca, Mg) CO_3.

DULCITOL n L *dulcis*, 'sweet', and **-ol**, denoting an alcohol (q.v.). A sweet crystalline substance, the hexahydric alcohol, $C_6H_8(OH)_6$, isomeric with mannitol, obtained from plants.

DURALUMIN n Perhaps fr L *durus*, 'hard', and **alumin(ium)**, but possibly from *Düren*, near Cologne, where it was first made. Proprietary name of a group of strong, light aluminium alloys.

DVI- Sanskrit, 'two' (cf. **eka-**). Numerical prefix employed by Mendeleef in 1871 to denote the second homologue of manganese, i.e. dvi-manganese, predicted by him on the basis of his periodic classification. (See **rhenium**).

Dy Chemical symbol for **dysprosium** (q.v.) the lanthanide element of at. no. 66.

DYNAMICS n From Gk δύναμις (dynamis), 'might, power'. The branch of physics which deals with the action of forces on bodies which results in motion, and also with the motion itself. (See **thermodynamics**.)

DYNAMITE n Gk δύναμις (dynamis), 'power'. Named coined by the inventor, Swedish chemist Alfred Nobel, in 1867, for his new explosive made by absorbing glyceryl trinitrate ('nitroglycerine') in kieselguhr.

DYNE n Gk δύναμις (dynamis), 'power'. Unit of force in the C.G.S. system, equal to 10^{-5} newton.

DYSPROSIUM n Gk δυσπρόσιτος (dysprositos), 'difficult to attain', fr δυσ- (dys-), 'difficult', πρός (pros), 'towards', ἰέναι (ienai) 'to go'. Name given to lanthanide element (at. no. 66) by Lecoq de Boisbaudran in 1886 in reference to the

difficulty experienced in isolating the element. Chemical symbol **Dy**.

DYSTECTIC adj Gk δυσ- (dys-) 'difficult', τηκτός (tektos), 'melted', i.e. 'difficult to melt'. Mixture of maximum melting point (cf. **eutectic**).

E

E Symbol for **energy**; symbol for electromotive force.

e Symbol for electronic charge.

EBULLIOSCOPY n Hybrid coined fr L *ebullire*, 'to boil up', and Gk σκοπεῖν (scopein), 'to look at, examine'. The study of boiling points; the determination of molecular weights by observing the elevation of boiling point of a solvent by a known concentration of solute (cf. cryoscopy).

EFFERVESCENCE n L *effervescere*, 'to boil over, ferment', fr *ex-*, 'out of, upward', and *fervescere*, 'to begin to boil or bubble'. Bubbling due to the formation of gas in a liquid.

EFFLORESCENCE n L *efflorescere*, 'to burst into flower', fr *ex-* 'out of, upward', and *florescere*, 'to come into blossom'. Change from crystals into fine powder through loss of water to the surrounding air.

52

EFFLUENT n L *effluere*, 'to flow out', fr *ex-* 'out of', *fluere*, 'to flow'. Waste liquid.

EICOSANE n Gk εἴκοσι (eicosi) 'twenty', and suff -ane (q.v.) denoting an alkane. The saturated hydrocarbon $C_{20}H_{42}$. (See Gk numerals Appendix 2.)

EIGENFUNCTION n Ger *Eigenfunktion*, 'proper function'. A solution of a partial differential equation (the Schrödinger equation) which has solutions only for particular values of the parameters.

EIGENVALUE n Ger *Eigenwert* 'proper value'. One of the particular values of the parameters of the Schrödinger equation for which the equation has a solution.

EINSTEINIUM After *Albert Einstein* (1879–1955). Name of transuranium element at. no. 99, proposed by the discoverers, Ghiorso, Thompson, Higgins, Seaborg *et al* in 1953 in honour of Einstein. Symbol Es.

EKA- Sanskrit *eka*, 'one'. Numerical prefix employed by Mendeléeff to denote the first homologues of boron, aluminium and silicon (i.e. eka-boron, eka-aluminium and eka-silicon), the existence of which he predicted on the basis of his periodic classification. These elements were later discovered (see scandium, gallium and germanium) (cf. **dvi-**).

ELAÏDIC adj Gk ἔλαιον (elaion), 'olive oil'. Elaïdic acid, $C_{17}H_{33}$-COOH, a stereoisomer of oleic acid, obtained from olive oil.

ELASTOMER n Abbreviation of *elastic polymer*, fr Gk ἐλαστικός

(elasticos) (fr ἐλα- (ela-), stem of ἐλαύνειν (elaynein), 'to drive') and **(poly)mer**. A polymer possessing elastic, rubbery properties.

ELECTRIC adj Gk ἤλεκτρον (electron), 'amber'. Term coined by William Gilbert in 1600 to denote those substances which, like amber, develop (electrical) attractive properties on being rubbed. The terms 'positive' electricity or 'negative' electricity were given by B. Franklin in 1747.

ELECTRO- Combining form meaning 'electrical'. See prec wd.

ELECTRODE n From **electro-** (q.v.) and Gk ὁδός (hodos), 'way'. Name coined by Faraday in 1834 to denote the part of an electrical circuit at which an electric current enters or leaves an electrolyte (cf. **anode**, **cathode**).

ELECTROKINETIC adj From **electro-** (q.v.) and **kinetic** (q.v.). In colloid chemistry, describes phenomena in which displacement of a disperse solid relative to a liquid medium occurs under the influence of an applied E.M.F., or in which an E.M.F. is generated by the displacement of the liquid medium and a solid disperse phase relative to one another.

ELECTROLYSIS n From **electro-** (q.v.), and **-lysis**, fr Gk λύσις (lysis), 'setting free'. Name coined by Faraday in 1834 to denote the chemical decomposition of substances in solution by an electric current.

ELECTROLYTE n See prec wd. A solution or melt which conducts an

53

electric current and is decomposed in the process.

ELECTRON n See **electric**. Name given by Johnstone Stoney in 1891 to the unit of electric charge carried by an ion, and by Wiechert in 1897 to the negatively charged particles constituting cathode rays. The elementary negatively-charged particle which is a constituent of all atoms.

ELECTRONICS n See prec wd. The branch of electrical science concerned with thermionic valves, cathode ray tubes, semiconductors etc., in which electrons move under control.

ELECTRO-OSMOSIS n From **electro-** (q.v.) and **osmosis** (q.v.). An electrokinetic (q.v.) phenomenon in which liquid moves through a membrane when an E.M.F. is applied to electrodes on opposite sides.

ELECTROPHILE n From **electron** (q.v.) and Gk φίλος (philos), 'loving'. A reagent which functions by accepting electrons (cf. **nucleophile**).

ELECTROPHORESIS n From **electro-** (q.v.) and Gk φορεῖν fr φέρειν (pherein), 'to carry, bear'. The migration of electrically charged colloid particles in an electric field.

ELECTROVALENCY n From **electro-** (q.v.) and **valency** (q.v.). Bonding due to electrostatic attraction between oppositely charged ions.

ELUATE n See **elution**. That which is eluted, e.g. the liquid emerging from the column in liquid chromatography.

ELUTE v See **elution**. To wash through with a solvent, as in liquid chromatography.

ELUTION n L *eluere*, 'to wash out'. Washing through of a column, as in chromatographic analysis.

EMETINE n Gk ἔμετος (emetos), 'vomiting'. An alkaloid, $C_{29}H_{40}N_2O_4$, extracted from the root of *Cephaelis ipecacuanha*, and so named because it provokes vomiting.

EMPIRICAL adj L *empiricus*, fr Gk ἐμπειρικός (empeiricos), 'experienced', fr ἐν (en), 'in', and πεῖρα (peira) 'experiment'. Based on experiment, practical.

EMULSION n L *emulsus*, 'milked out'. A dispersion of minute droplets of one liquid in another.

-EN- Shortened form of **-ene**, used in penultimate position of a name, e.g. 3-penten-2-one, $CH_3CH:CHCOCH_3$ to indicate presence of a double bond.

ENANTIOMER n Gk ἐναντίος (enantios), 'opposite', μέρος (meros), 'part'. An optical isomer; one of a pair of molecules which are mirror images of each other and are non-superimposable.

ENANTIOMORPHISM n Gk ἐναντίος (enantios), 'opposite', μορφή (morphe), 'shape'. Occurrence of a substance in two crystalline forms which are non-superposable mirror images of each other.

ENANTIOTROPY n Gk ἐναντίος (enantios), 'opposite', τρόπος (tropos), 'turn, direction'. The occurrence of a substance in two different physical forms, one stable above a certain temperature (the transition temperature) and the other stable below that temperature (e.g. rhombic and monoclinic sulphur).

ENDOCYCLIC adj Gk ἔνδον (endon), 'within', κύκλος (cyclos), 'circle'. Present within the ring (i.e. part of the ring) as distinct from **exocyclic**, outside the ring (i.e. projecting from the ring).

ENDOSMOSIS n Gk ἔνδον (endon), 'within', ὠσμός (osmos), 'push, impulse'. Inward flow of liquid through animal or vegetable membrane (Dutrochet, 1826).

ENDOTHERMIC adj Gk ἔνδον (endon), 'within', θέρμη (therme), 'heat'. A term introduced by Berthelot in 1879 to describe a chemical reaction in which heat is absorbed, or a compound which is formed from its elements with absorption of heat.

-ENE From **ethylene**. Name-ending proposed by Hofmann in 1866 to indicate presence of an ethylenic link, i.e. a double bond, in the molecule, as in butene, C_4H_8, and butadiene, C_4H_6.

ENERGY n L *energia*, fr Gk ἐνέργεια (energeia), fr ἐν (en), 'in', ἔργον (ergon) 'work'. Capacity for doing work, including potential, kinetic, electrical, chemical, nuclear energy.

ENOL n From **-ene** (q.v.) and suff **-ol** (q.v.) denoting an alcohol. Term introduced by Bruhl in 1894 to denote the form of a compound (such as acetoacetic ester) containing the group, $—CH=C(OH)—$, as distinct from the isomeric $—CH_2—CO—$ form, for which he used the term 'keto'.

ENSTATITE n Gk ἐνστάτης (enstates) 'opponent'. A pyroxene mineral, magnesium silicate, $MgSiO_3$, which derives its name from its refractory nature.

ENTHALPY n Gk ἐν (en), 'in', θάλπος (thalpos), 'heat'. Heat content. In thermodynamics, the quantity H which is equal to $U+pV$, where U is the internal energy, p the pressure and V the volume.

ENTROPY n Gk ἐν (en), 'in', and τροπή (trope), 'turning, change, transformation'. Lit 'transformation content'. Term introduced into thermodynamics by Clausius, on the analogy of the term energy. Entropy change is a measure of the irreversibility of a process; entropy may be regarded as the randomness factor of a system. Symbol **S**.

ENZYME n Gk ἐν (en), 'in', ζύμη (zyme), 'yeast'. Name given by Kühne in 1878 to some active agent, present in yeast, which effects fermentation. Any of a large group of protein substances able to bring about specific chemical reactions in living cells. Enzymes generally have names with the ending **-ase** following the name of the substrate, e.g. lipase, maltase, cellulase, amylase, or the reaction they bring about, e.g. coagulase, but some retain the name acquired before the adoption of this convention, e.g. emulsin, pepsin, trypsin.

EOSIN n Gk ἠώς (eos) 'dawn'. A rose-coloured dye, tetrabromofluorescein. $C_{20}H_8O_5Br_4$ (see **fluorescein, erythrosin**).

EPHEDRINE n From *Ephedra vulgaris*, botanical name of a desert shrub. An alkaloid, $C_6H_5CH(OH)$-$CH(CH_3)NHCH_3$, found in the plant *Ephedra* and used in medicine for raising the blood pressure and in the treatment of asthma and hay fever.

EPI- Combining form fr Gk ἐπί (epi) 'beside, on, after'. In chemistry, denotes a close relationship to the substance named.

EPOXY adj From Gk ἐπί (epi), 'beside', and **oxy(gen)**. Term used to denote compounds containing in the molecule an oxygen atom directly linked to two adjacent carbon atoms, thus forming a three-membered ring. Hence, epoxy resins, thermosetting resins made from 1-chloro-2,3-epoxypropane as one of the reactants.

EQUILIBRIUM n L *aequus*, 'equal', *libra*, 'balance'. Lit 'equal balance'. The condition in which there is no apparent change because the opposing influences are of equal force; e.g. in the isothermal equilibrium of a chemical system opposing reactions are occurring at identical rates so that no net change is observed.

Er Chemical symbol for **erbium**, lanthanide element, at. no. 68.

ERBIUM n From *Ytterby*, Sweden, the place from which came the mineral gadolinite in which Mosander discovered the element in 1843. The lanthanide element, at. no. 68, symbol **Er**.

ERG n Gk ἔργον (ergon) 'work'. The unit of work or energy in the C.G.S. system. 1 erg $= 10^{-7}$ joule.

ERGOSTEROL n From *ergot* and **sterol** (q.v.). A tetracyclic alcohol, $C_{28}H_{43}OH$, related to cholesterol, found in ergot and yeast and in small amounts in animal fats. UV irradiation of ergosterol produces calciferol, Vitamin D_2, necessary for healthy growth of bones and teeth in mammals.

ERYTHRITOL n Gk ἐρυθρός (erythros), 'red', and **-ol** (q.v.). A tetrahydric alcohol, CH_2OH-$(CHOH)_2CH_2OH$, derived from the colouring matter of certain lichens such as *Rocella tinctoria* and *Lecanora tinctoria* (from which litmus is obtained).

ERYTHRO- Combining form meaning 'red' or 'derived from some red material'. Gk ἐρυθρός (erythros), 'red'.

ERYTHROCYTE n Gk ἐρυθρός (erythros), 'red', (cytos), 'a hollow vessel'. Red blood cell, containing haemoglobin.

ERYTHROSE n Gk ἐρυθρός (erythros) 'red', and **-ose** (q.v.) signifying 'sugar'. A tetrose sugar $CH_2OH(CHOH)_2CHO$, closely related to **erythritol** (q.v.).

ERYTHROSIN n Gk ἐρυθρός (erythros), 'red', and **-in**, by analogy with **eosin** (q.v.). A red dye, tetraiodofluorescein, $C_{20}H_8O_5I_4$.

Es Chemical symbol for **einsteinium** (q.v.), transuranium element, at. no. 99.

-ESCENT Suffix meaning 'beginning to; beginning to be in the condition expressed by the verb'. L *-escens*, gen *-entis*, pres p suffix of verbs ending in -*escere*, meaning 'to begin to . . .'.

-ESCENCE Suffix forming abstract nouns from adjectives ending in -escent (q.v.).

ESSENTIAL OILS L *essentia*, 'essence of a thing'. Volatile oils distilled from plants and having the characteristic odour of the source plant.

ESTER n Ger *Ester*, fr Ger *Essig*, 'vinegar', and *Ather*, 'ether', name coined by Gmelin in 1848. Compound formed from an acid and an alcohol, e.g. ethyl acetate, CH_3-$COOC_2H_5$, the ethyl ester of acetic acid.

ETHANE n From eth(er) (q.v.) and -ane (q.v.). The second member of the paraffin series of saturated hydrocarbons, C_2H_6.

ETHANOL n From ethane (see prec wd) and suff -ol (q.v.) denoting an alcohol. Ethyl alcohol, C_2H_5OH, the product of fermentation of sugar and the active constituent of alcoholic beverages.

ETHER n L *aether*, fr Gk αἰθήρ (aither), 'pure upper air' (cf. Persian '*attar*'). Name first used in its modern sense by Frobenius in 1730 for the substance he named 'spiritus aethereus'. Compound in which two hydrocarbon radicals, which may be the same or different, are joined to an oxygen atom e.g. diethylether, $(C_2H_5)_2O$, and methylethylether $CH_3OC_2H_5$.

ETHYL adj From Gk αἰθήρ (aither) 'pure upper air', ὕλη (hyle) 'stuff'. Name coined by Liebig in 1840 to denote the radical common to alcohol, ether and certain esters. The saturated hydrocarbon radical C_2H_5.

ETHYLENE n From ethyl (q.v.) and -ene, the word-termination proposed by Hofmann in 1866 for unsaturated compounds. The olefin C_2H_4.

ETYMOLOGY n F *etymologie*, fr L *etymologia*, fr Gk ἐτυμολογία (etymologia), fr ἔτυμον (etymon) 'the true sense of a word according to its origins', and -λογία (-logia) 'study', hence, the study of the true sense of words in the light of a knowledge of their origins.

Eu Chemical symbol for europium (q.v.) lanthanide element, at. no. 63.

EUDIOMETER n Gk εὔδιος (eudios), 'fine', μέτρον (metron), 'measure'. Name coined by Priestley in 1777 for his apparatus for determining the 'fineness' of air, i.e. the proportion of oxygen contained in it.

EUGENOL n From *Eugenia caryophyllata*, botanical name of the clove tree, and (phen)ol (q.v.). A phenol $C_6H_3(OH)(OCH_3)CH_2CH$:CH_2, found in oil of cloves.

EUROPIUM n From *Europe* and -ium (q.v.). Name given in honour of Europe to the lanthanide element, at. no. 63, by Demarçay, who separated the element from samarium in 1896.

EUTECTIC n or adj Gk εὖ (eu), 'easily', τήκειν (tekein), 'to melt'. A mixture of two or more substances, possessing the lowest melting point of all possible mixtures of them.

EUTROPHICATION n Gk $\epsilon\hat{v}$ (eu) 'well', $\tau\rho\acute{\epsilon}\phi\epsilon\iota\nu$ (trephein), 'to nourish'. The supplying of excessive amounts of nutrient material to the environment, with consequential upsetting of the natural balance.

EXOCYCLIC adj Gk $\acute{\epsilon}\xi\omega$ (exo), 'outside', $\kappa\acute{v}\kappa\lambda os$ (cyclos), 'circle, ring'. Outside the ring, e.g. the oxygen atom in cyclohexanone as distinct from the **endocyclic** (q.v.) oxygen atoms in dioxan.

EXOTHERMIC adj Gk $\acute{\epsilon}\xi\omega$ (exo), 'out', $\theta\acute{\epsilon}\rho\mu\eta$ (therme), 'heat, giving out heat'. The term was introduced by Berthelot in 1879 (cf. **endothermic**).

EXPERIMENT n or v L *experimentum*, 'trial, test', fr L *experiri*, 'to try'. Practical trial or test.

EXPLODE v L *explaudere*, 'to drive off the stage by hooting, whistling or clapping'. To disrupt violently as a result of sudden development of pressure.

EXTRACT v or n L *extractus*, 'having been drawn out', fr L *ex*, 'out' and *trahere*, 'to draw'. To draw out of a solid or out of a solution, as in solvent extraction.

F

F Chemical symbol for **fluorine**, the halogen element, at. no. 9. Symbol for the **farad**, the unit of capacitance, also for Faraday's constant, 9·64870 C/mol.

f Symbol for **femto-** (q.v.) signifying 10^{-15} in the SI system. Symbol for angular quantum number three (arising indirectly from the 'fundamental' series in the atomic spectra of the alkali metals). Symbol for **fugacity**.

FARAD n After *Michael Faraday* (1791–1867). Unit of electrical capacitance. Symbol F.

FARNESOL n From *Farnesiana*, a species of acacia (named after Cardinal Farnese (1573–1626)) in which it occurs. A sesquiterpene alcohol, $C_{15}H_{25}OH$, found in ambrette seed oil, and in lime and cyclamen flowers.

FAT n OE *faett*. The semi-solid oily or greasy substance present in the animal body and consisting of glyceryl esters of fatty acids.

FATTY ACIDS From **fat** (q.v.) and **acid** (q.v.). Members of the homologous series of alkanoic acids $C_nH_{2n+1}COOH$, e.g. acetic acid, CH_3COOH, stearic acid, $C_{17}H_{35}COOH$, so called because many of them are present in fats in the form of glyceryl esters.

Fe Chemical symbol for **iron** (from L *ferrum*, 'iron') element at. no. 26.

FELDSPAR See **felspar**.

FELSPAR n Ger *Feldspat(h)*, fr *Feld*, 'field', and *Spat(h)*, 'spar, crystal-line mineral', so called because the mineral occurs in some areas in the form of large loose crystals lying about in the field and detached from the rock in which they were originally included. The more common form 'felspar' is, strictly, incorrect. A group of crystalline aluminosilicate minerals, e.g. orthoclase, $KAlSi_3O_8$.

FEMTO- Prefix in the SI system of units (see Appendix 8) denoting 10^{15}. (cf. Danish *femten*, Swedish *femton*, 'fifteen'.) Symbol **f**.

FENCHONE n Ger *Fenchel*, 'fennel', **-one**, 'ketone'. A terpene ketone, $C_{10}H_{16}O$, found in fennel.

FERMENTATION n L *fermentare* 'to cause to rise', fr L *fervere*, 'to boil'. The process of conversion of certain naturally occurring organic compounds into simpler products by enzymes, e.g. the conversion of sugar into alcohol and carbon dioxide by the action of yeast.

FERMIUM n After *Enrico Fermi* (1901–1954). The radioactive transuranium element at. no. 100, named in honour of Fermi by its discoverers Ghiorso, Thompson, Higgins, Seaborg and others in 1953. Symbol **Fm**.

FERRIC adj L *ferrum*, 'iron'. Obsolescent term denoting compound of tervalent iron, i.e. of iron (III), e.g. ferric chloride, $FeCl_3$.

FERRIFEROUS adj L *ferrum*, 'iron', *fero*, 'I bear'. Iron-bearing, in reference to minerals and rocks.

FERROUS adj L *ferrum*, 'iron'. Obsolescent name for compounds of bivalent iron, i.e. of iron (II), e.g. ferrous sulphate, $FeSO_4$.

FILTER n or v Med L *feltrum*, 'felt'. A device for separating solid matter from liquids by passage through a porous material.

FISSION n L *fissio*, gen *fissionis*, 'splitting'. Cleaving or splitting, e.g. of the nucleus of a heavy element such as uranium or plutonium, into two roughly equal parts.

FLAVINE n L *flavus*, 'yellow'. A yellow acridine dye.

FLAVONE n Ger *Flavon*, fr L *flavus*, 'yellow'. The substance 2-phenylchromone, $C_{15}H_{10}O_2$, (which is colourless) or any of its derivatives, many of which occur in nature as plant pigments.

FLOCCULATE v L *flocculus* 'a little flock of wool'. To coagulate into little wool-like lumps.

FLUORENE n From **fluor(spar)** on account of the fluorescence of the compound when impure. The aromatic hydrocarbon, $C_{13}H_{10}$, o-diphenylenemethane.

FLUORESCEIN n From fluoresce (see foll wd). A phthalein dye, $C_{20}H_{12}O_5$, made by condensation of resorcinol and phthalic anhydride, which fluoresces intensely in alkaline solution.

FLUORESCENCE n From **fluor-(spar)** (q.v.) by analogy with opalescence (see **opalescent**). The property of fluorspar and many other substances, e.g. fluorescein and quinine sulphate, of showing opalescence of colour different from that of the incident light.

FLUORIDATION n From **fluoride** (q.v.). The addition of minute amounts (ca. 1 ppm) of fluoride to drinking water with the object of giving protection against dental caries.

FLUORIDE n From **fluor(ine)** (q.v.). Compound of fluorine with another element; salt of hydrofluoric acid, e.g. calcium fluoride, CaF_2.

FLUORINE n From **fluor(spar)** and **-ine**, by analogy with chlorine; name given by Davy in 1813, following a suggestion by Ampère, to the element, analogous to chlorine, present in fluorspar. The halogen element, at. no. 9; symbol F.

FLUORITE n From **fluorspar**. Alternative name of fluorspar, mineral calcium fluoride.

FLUORO- From **fluor(ine)**. Fluorine-containing, e.g. fluorocarbons, fluorinated hydrocarbons.

FLUORSPAR n L *fluor*, 'a flow, flowing' and Ger *Spat(h)*, 'spar, crystalline mineral', so called on account of its crystallinity and ready fusibility. Mineral calcium fluoride, CaF_2.

FLUX n L *fluere*, 'to flow'. A substance added to promote fusion in metallurgical processes. Flow, as in neutron flux, the number of neutrons flowing through 1 cm^2 in any direction in one second.

Fm Chemical symbol for **fermium** (q.v.) transuranium element, at. no. 100.

FOLIC adj L *folium*, 'leaf'. Folic acid, pteroylglutamic acid, $C_{19}H_{19}N_7O_6$, a vitamin of the B group found in green leaves and used in the treatment of anaemia.

FORENSIC adj L *forensis*, 'concerned with the forum', i.e. to do with the law, as Roman law courts were usually sited near the forum. Forensic chemistry is chemistry concerned with matters of law.

FORMALDEHYDE n From **formic acid**, (q.v.) and **aldehyde** (q.v.), 'the aldehyde which yields formic acid on oxidation'. The first member of the homologous series of aliphatic aldehydes, HCHO, a pungent gas readily soluble in water yielding a 40% solution known as **formalin**.

FORMALIN n From **formal-(dehyde)** (q.v.). A 40% solution of formaldehyde in water.

FORMIC adj L *formica*, 'ant'. Formic acid, HCOOH, so named because first obtained, in 1749, by distillation of red ants.

Fr Chemical symbol for **fr(ancium)** (q.v.), radioactive alkali metal, at. no. 87.

FRANCIUM n Mod L *Francia*, 'France', the native land of Marguerite Perey, who discovered the element in 1946. The radioactive alkali metal, element at. no. 87, symbol **Fr**.

FRUCTOSE n L *fructus*, 'fruit', and suff **-ose** (q.v.) denoting 'sugar'. A sugar, $C_6H_{12}O_6$, found in many sweet fruits and in honey. It is a laevorotatory keto-hexose and is produced, together with glucose, on hydrolysis of sucrose.

FUCHSIN n From *Fuchsia*, the ornamental shrub (which is named after *Leonard Fuchs*). A triphenylmethane dye (rosaniline) of vivid red colour resembling that of the flower fuchsia.

FUGACITY n L *fugax*, gen. *fugacis*, 'apt to flee, transitory'. A thermodynamic quantity, symbol f, related to gas pressure P, such that f/P tends to unity as P tends to zero.

FULLER'S EARTH n From Old F, *fuler*, 'to cleanse and thicken cloth'. Any earthy substance (aluminosilicate) which can be used to decolorise mineral or vegetable oils, as in fulling cloth.

FULMINATE n L *fulminare*, 'to strike with lightning'. Salt of fulminic acid, HONC (an isomer of cyanic acid, HOCN), e.g. mercury (II) fulminate, used as a detonator.

FUMARIC adj From *Fumaria*, botanical name of fumitory, the plant in which fumaric acid is found. Fumaric acid, the unsaturated dicarboxylic acid, HOOCCH: CHCOOH, geometric isomer of maleic acid, of which it is the trans-form.

FURAL n L *furfur*, 'bran', suff -al, denoting an aldehyde. The aldehyde $C_5H_4O_2$, of furan (q.v.), so named because it was obtained, by Fownes in 1845, by distillation of bran with dilute sulphuric acid.

FURAN n L *furfur* 'bran'. Abbreviation of **furfuran**. The five-membered heterocyclic compound $(CH)_4O$.

FURANOSE adj From **furan** (q.v.) and **-ose** (q.v.) denoting a sugar. Sugar structure featuring a five-membered ring of the same kind as that in **furan** (q.v.) (as distinct from pyranose, in which there is a six-membered ring like that of **pyran**).

FURFURAL n L *furfur*, 'bran'. Same as **fural** (q.v.).

FURFURAN n L *furfur*, 'bran'. Same as **furan** (q.v.).

FURFURYL adj L *furfur*, 'bran'. Furfuryl alcohol $C_4H_3OCH_2OH$, the alcohol formed by reduction of the aldehyde group in fural.

FUSEL OIL Ger *Fusel*, 'bad brandy', fr Ger *fuseln*, 'to bungle'. The mixture of several homologous alcohols, mostly amyl alcohol, produced as a by-product during alcoholic fermentations. Of unpleasant odour and taste, it is removed in the production of potable spirits by distillation.

-FY Suffix denoting 'to make into'. L *facere*, 'to make, do'.

G

G Symbol for **Giga** (q.v.), prefix signifying 10^9 times the fundamental unit in the SI system. Symbol for Gibbs free energy, also for **Gauss**, the unit of magnetic induction.

g Symbol for gravitational acceleration.

Ga Chemical symbol for **gallium** (q.v.), element at. no. 31.

GADOLINITE n After *Johan Gadolin* (1760–1852), mineralogist and discoverer of the mineral. A

silicate mineral found in Scandinavia
and source of yttrium, cerium,
lanthanum and several lanthanide
elements.

GADOLINIUM n After *Johan
Gadolin* (1760–1852); name given by
the discoverer of the element, Lecoq
de Boisbaudran in 1886. The lanthanide element, at. no. 64, symbol
Gd.

GALACTOSE n Gk γάλα (gala),
gen γάλακτος (galactos), 'milk', and
-ose (q.v.), denoting 'a sugar'. The
hexose sugar, CHO(CHOH)₄CH₂-
OH, present in milk, discovered by
Pasteur in 1856 and given its present
name by Berthelot in 1860.

GALENA n L *galena*, name used
by Pliny for the dross left after lead
smelting. Mineral lead sulphide,
PbS.

GALLIC adj F *gallique*, fr L *galla*,
'gall nut'. Gallic acid, 3,4,5-tri-
hydroxybenzoic acid, found in gall
nuts.

GALLIUM n L *Gallia*, the old
name of France. The metallic element
(Mendeléeff's eka-aluminium) discovered in 1875 by Lecoq de Boisbaudran and named in honour of his
native country. Element at. no. 31,
symbol **Ga**.

GALVANISE v After *Luigi
Galvani* (1737–1798), early investiga-
tor of electricity. To cover with metal
by electrodeposition. The term is
used incorrectly when applied to the
coating of steel with zinc by other
methods.

GAMMA RADIATION n Gk
γάμμα (gamma), the third letter of

the Gk alphabet. The third type of
radiation observed in early studies
of radioactivity. Penetrating electromagnetic radiation of the same
nature as X-rays but of shorter
wavelength and higher energy; emitted by nuclei in radioactive decay.

GANGUE n F *gangue*, fr Ger
Gang, 'a vein of metal in rocks'. The
clayey and stony matter intimately
associated with metallic ore deposits.

GARNET n OF *grenate*, fr L
granatum, 'pomegranate', possibly so
named because the colour of the
mineral resembles that of the crushed
fruit. Group of hard crystalline
mineral silicates, some of which are
red and used as gems.

GAS n A word first used by
Paracelsus (c. 1493–1541) but first
employed in its modern sense by van
Helmont (1577–1644) and said by
him to have been suggested by
Gk χάος (chaos), 'formless void'. A
substance in its least dense condition,
in which state its constituent molecules are separated by substantially
greater distances than in the liquid
or solid states.

GASEOUS adj See prec wd.

GAUSS n After *Gauss* (1777–
1855), who strongly advocated the
expressing of all physical quantities
in terms of mass, length and time.
The C.G.S. unit of magnetic flux
density.

Gd Chemical symbol for **gadolinium** (q.v.) the lanthanide element, at.
no. 64.

Ge chemical symbol for **germanium** (q.v.) element at. no. 32.

GEL n From **gel(atin)** (q.v.). A colloidal system in the state of a jelly and consisting of an open framework of disperse phase with the intervening space filled with the dispersion medium.

GELATIN n F *gélatine*, fr Ital *gelatina*, fr L *gelare*, 'to freeze, to set solid'. The protein resulting from the hydrolysis of collagen in bones and animal tissues on boiling with water.

GELATION n See prec wds. The process of setting of a gel.

GEM n L *gemma*, 'precious stone, bud'. Jewel, precious stone.

GEM- Abbreviation of **geminal** (see foll wd). Chemical prefix applied to two like substituents on the same carbon atom (cf **vic-**), e.g. 1,1-dichlorethane, CH_3CHCl_2, is gem-dichlorethane, whereas 1,2-dichlorethane, CH_2ClCH_2Cl, is vic-dichlorethane.

GEMINAL adj L *gemini*, 'twins'. Term applied in organic chemistry to a pair of substituent groups attached to the same carbon atom in a chain; abbreviation **gem-** (see prec wd); cf. **vicinal**.

GEOCHEMISTRY n Gk γεω- (geo-) 'earth' (fr γῆ (ge) 'earth'), and 'chemistry' (q.v.). The branch of chemistry concerned with the composition of the earth's crust and the chemical processes occurring within it.

GERANIAL n See foll wd. A terpene aldehyde $C_9H_{15}CHO$, corresponding to **geraniol**, found in lemon grass oil and geranium oil, and used in perfumery.

GERANIOL n From *geranium* (fr Gk γέρανος (geranos) 'crane', geranium being a genus of plants bearing a fruit resembling the bill of a crane in shape). A terpene alcohol $C_9H_{15}CH_2OH$, a constituent of many essential oils and the chief constituent of Indian geranium oil, much used in perfumery.

GERMANIUM n L *Germania*, the Latin name of the country of its discoverer, Winkler. The element predicted under the name of ekasilicon by Mendeléeff in 1871 and discovered by Winkler in 1886. At. no. 32, symbol Ge.

GIBBSITE n Named after *George Gibbs*, American mineralogist in 1822. A form of aluminium trihydrate, $Al(OH)_3$.

GIBERELLIN n From *Giberella*, name of a fungus. A group of substances obtained from cultures of the fungus *Giberella* and used as plant growth regulators.

GIGA- Gk γίγας (gigas) 'giant' (cf. 'gigantic'). Prefix in SI system denoting 10^9 times the fundamental unit (see **SI Units**). Symbol **G**.

GLASS n ME *glas*, OE *glaes*, OF *glaes*. A supercooled melt of silicates, or other substances, in which the immediate environment of the atoms or molecules may be similar to that in the crystalline state but in which long-distance order is absent.

GLUCOSE n From Gk γλυκύς (glycys), 'sweet'. Grape sugar, $C_6H_{12}O_6$, an aldo-hexose, present in sweet fruits and in honey, dextro-rotatory (hence the alternative name dextrose), produced by hydrolysis of cellulose and starch, and important as a foodstuff.

GLUCOSIDE n See prec wd. A compound of glucose (or some other sugar) with another substance (not a sugar) which on hydrolysis yields the sugar and the other substance (which is termed an 'aglycone'). Glucosides occur very widely in nature (see, for example, **amygdalin** and **anthocyanidin**).

GLUTAMIC adj Ger *Glutaminsäure*. See **gluten**, and 'amine'. **Glutamic acid**, α-aminoglutaric acid, $COOH(CH_2)_2 CH(NH_2)COOH$, (originally named 'amidogen') occurs in vegetable protein and is used, in the form of its sodium salt, as a condiment.

GLUTAMINE n See prec wd. The aminoacid $COOH CH(NH_2) (CH_2)_2 CONH_2$, the half-amide of glutamic acid.

GLUTARIC adj From **gluten** (q.v.), and **tartaric** acid. Glutaric acid, the dicarboxylic acid $COOH(CH_2)_3COOH$.

GLUTEN n L *gluten*, 'glue'. A mixture of proteins present in wheat flour obtained as a very sticky yellowish mass by making a dough and washing out the starch.

GLYCERIC adj From **glycerol** (q.v.). Glyceric acid, $HOCH_2 CHOHCOOH$, the acid obtained by oxidation of glycerol.

GLYCERIDE n From **glycer(ol)** (q.v.) and **-ide** (q.v.). Ester of glycerol with an organic acid. Animal fats and vegetable oils are glycerides of long-chain saturated fatty acids, e.g.

stearic acid, $C_{17}H_{35}COOH$, and unsaturated fatty acids, e.g. linoleic acid, $C_{17}H_{31}COOH$, respectively. On hydrolysis they yield glycerol and the acids.

GLYCEROL n Gk γλυκερός (glyceros), 'sweet', and **-ol** (q.v.), denoting alcohol. The trihydric alcohol, propan-1,2,3-triol, $HOCH_2$-$CHOHCH_2OH$, a colourless, viscous, sweet liquid, a constituent of animal fats and vegetable oils.

GLYCERYL adj From **glycer(ol)** (q.v.) and **-yl** (q.v.). The radical C_3H_5 derived from glycerol and present in glycerides.

GLYCINE n Gk γλυκύς (glycys), 'sweet'. Name coined by Berzelius in 1848 for the sweet crystalline substance which had been discovered in 1820 by Braconnot by boiling glue with dilute acid, and called 'sugar of gelatin'. The simplest α-aminoacid, aminoacetic acid, NH_2CH_2COOH, also known as **glycocoll** (q.v.), a general constituent of proteins.

GLYCOCOLL n Gk γλυκύς (glycys), 'sweet', κόλλα (colla) 'glue'. Name proposed by Horsford in 1846 for the substance now known as **glycine** (see prec wd).

GLYCOGEN n Gk γλυκύς (glycys), 'sweet', and *-gen*, 'producer', fr Gk γείνομαι (geinomai), 'I produce'. Animal starch, a polysaccharide, $(C_6H_{10}O_5)_n$, found in the animal liver and other organs, a supplier of sugar in the body (hence the name).

GLYCOL n From F *glycérine*, glycerol (q.v.) and F *alcool*, 'alcohol'. Name given by Wurtz in 1856 to 'mark the double analogy between this compound with glycerol on the one hand, and alcohol on the other'. The compound ethylene glycol, ethan-1,2-diol, $HOCH_2CH_2OH$, a sweet viscous liquid.

GLYCOLLIC adj From **glycocoll** (q.v.). Glycollic acid, $HOCH_2COOH$, hydroxyacetic acid, obtained by Strecker in 1848 from glycocoll (i.e. glycine) by the action of nitrous acid.

GLYCOSIDE n Gk γλυκύς (glycys) 'sweet', and suff -**ide**. General name for compounds of a sugar (e.g. glucose, galactose, mannose) with another substance (not a sugar, and called an 'aglycone'). (See **glucoside**.)

GLYOXAL n From **gly(col)** and **oxal(lic) acid.** The compound CHOCHO, intermediate between glycol CH_2OHCH_2OH and oxalic acid COOHCOOH.

GLYOXYLIC adj From **gly(col)** and **ox(alic) acid.** The compound, CHO.COOH, intermediate between glyoxal and oxalic acid.

GNEISS n Ger *Gneiss*, 'sparkling'. A rock containing irregular bands of quartz, felspar and mica; a metamorphosed granitic rock.

GOLD n ME *gold*, fr OE *gold*, 'gold'. The precious yellow metal, element 79. Symbol **Au**, from L *aurum*, 'gold'.

-GRAM Combining form denoting something written or drawn, as in *diagram, chromatogram*. From Gk γράμμα (gramma), 'that which is written'.

GRAMME n F *gramme*, fr Late L *gramma*, fr Gk γράμμα (gramma), 'a letter, a small weight'. The fundamental unit of mass in the C.G.S. system of units, 10^{-3} kg.

GRANITE n Ital *granito*, 'grained, granular', fr L *granum*, 'grain', so named on account of its grainy appearance. A hard, durable, crystalline, igneous rock consisting of felspar, mica and quartz.

GRAPH n (also v) Gk γράφειν (graphein), 'to write, draw, inscribe'. A curve representing the mathematical relationship between two quantities, in particular those obtained experimentally.

-GRAPH Combining form denoting (1) something that is written or drawn, (2) that which writes or portrays. From Gk γράφειν (graphein), 'to write', 'to write, draw, inscribe'.

GRAPHITE n Ger *Graphit* fr Gk γράφειν (graphein), 'to write', so named in 1789 because of its use in making pencils. A soft, black, flaky mineral consisting of an allotrope of carbon.

GRISEOFULVIN n From *Penicillium griseofulvum* (fr Med L *griseus* 'grey', *fulvus*, 'yellowish brown') a species of *Penicillium* which yields this substance. The compound $C_{17}H_{17}O_6Cl$, used as an antibiotic and in the treatment of fungus diseases.

GUAIACOL n From *Guaiacum* (name of a tree growing in the West Indies, and of the resin obtained from it) and (**phen**)**ol**. A phenol, 2-methoxyphenol, obtained by the dry distillation of guaiacum resin.

GUANIDINE n From **guanine** (see foll wd). A water-soluble, alkaline organic substance, $HN:C(NH_2)_2$, obtained from guanine by oxidation.

GUANINE n From **guano** (see foll wd). A compound of the purine group, 2-aminohypoxanthine, $C_5H_5N_5O$, found by Unger in 1846 in guano.

GUANO n Spanish, *guano*, fr S American word '*huann*, 'dung'. A natural manure consisting largely of the accumulated excrement of seabirds, found in S America and particularly on islands off Peru, and valued as a fertiliser.

GUNMETAL n From gun and '**metal**'. An alloy of copper and tin formerly important in gunmaking.

GUTTA PERCHA n From Malay wds *gutta*, 'gum' and *percha*, the tree from which it is obtained. A kind of thermoplastic rubber from Malaya, the trans-form of polyisoprene.

GYPSUM n L *gypsum*, fr Gk γύψος (gypsos) 'chalk, gypsum'. Calcium sulphate dihydrate, $CaSO_4\cdot2H_2O$, from which plaster of Paris may be made by heating to 120° C, when 75% of the water is driven off.

H

H Chemical symbol for **hydrogen** (q.v.), element of at. no. 1.

h Symbol for the Planck constant (quantum theory). Symbol for **hecto-** (q.v.) denoting 10^2 in the SI system.

HAEM n Gk αἷμα (haima), 'blood'. The non-protein portion of the haemoglobin molecule (q.v.); formula $C_{34}H_{32}O_4N_4Fe^{II}$.

HAEMATIN or **HAEMATINE** n Formed with suff **-in** or **-ine**, fr Gk αἷμα (haima), gen αἵματος (haimatos), 'blood'. The colouring matter of haemoglobin; formula $C_{34}H_{33}O_5N_4Fe^{III}$.

HAEMATITE n L *haematites*, fr Gk αἱματίτης (haimatites), 'blood-like', fr αἷμα (haima), gen αἱμάτος (haimatos), 'blood'. Mineral form of Fe_2O_3, some of the massive varieties of which are blood red. The name dates from the time of Theophrastus about 325 B.C.

HAFNIUM n L *Hafnia*, 'Copenhagen' (cf. Danish København). Metallic element, at. no. 72, named by its discoverers Coster and Hevesy in 1923 in honour of Copenhagen. Symbol Hf.

HALOGEN n Gk ἅλς (hals), gen ἁλός (halos), 'salt', and γεννᾶν (gennan) 'to produce'. Lit 'salt-producer', named coined by Berzelius to denote any one of the four elements, **fluorine, chlorine, bromine, iodine**.

HASHISH n Arab *hashish*, 'hemp', properly 'dried grass', related to Hebrew *háshash*, 'dried grass'. The top leaves and tender parts of Indian hemp, dried for smoking or chewing.

He Chemical symbol for **helium** (q.v.), element of at. no. 2.

HECTO- (before a vowel **hect-**) Combining form meaning 'a hundred'. F *hecto, hect,* from Gk ἑκατόν (hecaton) 'a hundred'.

HELIUM n Gk ἥλιος (helios), 'sun'. The noble gas, element of at. no. 2, so named by Sir Norman Lockyer and Sir Edward Frankland, because its existence was first inferred by examination of a solar spectrum, obtained during the eclipse of 1868. The element was first discovered on earth by Sir William Ramsay in 1895 in the gas evolved from the mineral cleveite. Symbol **He.**

HEMATIN or **HEMATINE** n Synonymous with **haematin** (q.v.).

HEMATITE n See **haematite.**

HEMI Prefix meaning 'half'. Gk ἡμι- (hemi-), cognate with L *semi,* 'half'.

HENRY n The unit of inductance. Named after the American physicist *Joseph Henry* (1797–1878).

HEPTA (before a vowel **HEPT**) Combining form meaning 'seven'. Gk ἑπτά (hepta), 'seven'.

HEROIN also **HEROINE** n Ger *Heroin,* coined by H. Dreser, probably from Gk ἥρως (heros) 'hero' and chem suff **in** or **ine.** An addictive drug obtained from morphine.

HERTZ n The unit of frequency in the SI system. Symbol Hz.

HETERO (before a vowel **HETER-**) Combining form meaning 'other, different'. Gk ἑτερο-, ἑτερ-, fr ἕτερος (heteros), 'the other (of two); another, different'.

HETEROCYCLIC adj From **hetero-** (see prec entry) and **cyclic** (q.v.). Heterocyclic compounds are normally organic compounds which contain a closed ring system in which the atoms are of more than one kind, e.g. pyridine, furan.

HETEROGENEOUS adj Med L *heterogeneus,* fr Gk ἑτερογενής (heterogenes), 'of different kind', fr ἕτερος (heteros) 'another, different', and γένος (genos), 'race, gender, kind'. Not homogeneous; disparate.

HETEROPOLY adj From **hetero-** (q.v.) and **poly-** (q.v.). As in heteropoly acids and their salts, a group of complex acids which contain at least two different acid radicals, one of them usually in large numbers, e.g. tungstophosphonic acid, $H_3(P(W_3O_{10})_4)$.

HEULANDITE n A mineral named by Brooke in 1822 in honour of *H. Heuland,* an English mineral collector. $H_4CaAl_2(SiO_3)_6 . 3H_2O$.

HEURISTIC adj Ger *heuristisch* from Mod L, *heuristicus,* which is formed with the suff *-isticus* from the stem of Gk εὑρίσκειν (heuriskein), 'to find, discover'. Serving to discover or find out.

HEXA- (before a vowel **HEX-**) Combining form meaning 'six'. Gk ἑξα-, ἑξ- (hexa-, hex-), 'six'.

Hf Chemical symbol for **hafnium** (q.v.), element at. no. 72.

Hg Chemical symbol for **mercury** (q.v.) (fr L *hydrargyrum,* 'quicksilver, mercury'), element at. no. 80.

HIPPURIC adj From Gk ἵππος (hippos), 'horse', and οὖρον (ouron) urine'. Hippuric acid, $C_6H_5CO-NHCH_2COOH$, first found in the urine of horses.

HISTAMINE n From **histo-** (q.v.) and **amine**. 4-aminoethylglyoxaline $C_5H_9N_3$, found in many animal tissues and also in ergot.

HISTIDINE n From **hist(o)-** (q.v.) and **(am)ine**. α-amino-β-imidazolyl propionic acid, $C_6H_9O_2N_3$, occurs in animal tissues.

HISTO- (before a vowel **HIST-**) Combining form meaning 'tissue'. Gk ἱστός (histos), 'web'.

Ho Chemical symbol for **holmium** (q.v.), element at. no. 67.

HOLMIUM n From *Holmia*, ancient name of Stockholm. The rare lanthanide element of at. no. 67, discovered in 1879 by Cleve and so named in honour of Stockholm, his native city. Symbol **Ho**.

HOMO- (before a vowel **HOM-**) Combining form meaning 'one and the same, jointly'. Gk ὁμο- (homo-), ὁμ- (hom-), fr ὁμός (homos), 'one and the same, jointly'.

HOMOGENEOUS adj Med L *homogeneus*, fr Gk ὁμογενής (homogenes) 'of the same kind or race', fr ὁμο- (homo-), and γένος (genos), 'race, gender, kind'. Of the same kind.

HOMOLOGOUS adj Gk ὁμόλογος (homologos), 'agreeing, of one mind'. Compounded of ὁμο- (homo-) (q.v.) and λόγος (logos) 'word, speech'. In org chem a homo-

logous series is one in which the successive members differ in composition by a constant amount of certain constituents (esp CH_2) and show a gradation of chem and phys properties.

HORMONE n Gk ὁρμῶν (hormon) 'that which urges or arouses', pres part ὁρμᾶν (horman), 'to set in motion, urge, stimulate'. Used by Hippocrates to denote a vital principle. Endocrine gland secretion which stimulates (physiological) activity. (First used in modern sense by Starling in 1903.)

HUMUS n L *earth, ground, soil*. The characteristic organic constituent of soils.

HYBRID n L *hybrida, hibrida*, 'mongrel'. (1) In chem, a molecule which results when electronic orbitals of differing types but similar energies become combined to form equivalent orbitals, but without change in the total number of orbitals, e.g. one s and one p orbital can give two sp orbitals. (2) In linguistics, a word compounded of elements from different languages.

HYDRATE n A hybrid term coined by J. L. Proust about 1800 fr Gk ὕδωρ (hydor), 'water' (see **hydro-**) and **-ate**, a chem suff of Latin origin. A compound in which one or more molecules of water are attached to an ion in solution, or are present as such in the lattice of a crystalline compound.

HYDRAZINE n Coined from **hydr(ogen)**, **azo-** (denoting the presence of nitrogen) and the suff **-ine**. The compound N_2H_4.

68

HYDRAZOIC adj Coined from **hydr(ogen)**, **azo-** denoting presence of nitrogen, and suff **-ic**. Hydrazoic acid is HN_3.

HYDRAZONE n From **hydrazine** by change of letter. Name coined by E. Fischer in 1888 for compounds of aldehydes and ketones with hydrazine or a substituted hydrazine. "The letter 'o' immediately reminds one of the azo compounds, with which the hydrazones are isomeric".

HYDRO- (before a vowel **HYDR-**) Combining form denoting: (i) the presence of hydrogen; (ii) water. Gk ὕδωρ (hydor), 'water'.

HYDROCARBON n A compound composed of **carbon** and **hydrogen** only.

HYDROGEN n F *hydrogène*, fr fr Gk ὕδωρ (hydor), 'water', and γείνομαι (geinomai), 'I produce'. Name coined by Lavoisier in 1789 in reference to the formation of water on combustion of this element. Element of at. no. 1, symbol **H**.

HYDROGENATE v From **hydrogen** and suff *-ate*. To add hydrogen to a compound; to cause a compound to combine with hydrogen.

HYDROLYSIS n Compound of **hydro-** (q.v.) and Gk λύσις (lysis) 'a loosening, setting free'. Decomposition of a compound by the action of water.

HYDROPHILIC adj Gk ὕδωρ (hydor) 'water' and φίλος (philos), 'loving'. Lit 'water-loving'. Applied in colloid chem to denote a disperse phase (e.g. gelatin) with a high affinity for water (cf. the more general term **lyophilic**, q.v.). First adopted by M. Perrin in 1905, the term is now applied in surface chem to refer also to surfaces which are readily wetted by water.

HYDROPHOBIC adj Gk ὕδωρ (hydor) 'water', and φόβος (phobos) 'fear'. Lit 'water-fearing'. Applied in colloid chemistry to denote a disperse phase (e.g. arsenic sulphide) with a low affinity for water. (Cf. the more general term **lyophobic**, q.v.). First adopted by M. Perrin in 1905, the term is now applied in surface chem to refer also to surfaces which are water-repellent.

HYDROSTATIC adj Compounded of **hydro-** and Gk στατίκος (statikos), 'causing to stand' fr στατός (statos), 'placed, standing'. Relating to the equilibrium of liquids and to the pressure exerted by liquids at rest.

HYDROXY- Combining form indicating the presence of a **hydroxyl** radical. Formed from **hydr(ogen)** and **oxy(gen)**.

HYDROXYL n Formed from **hydr(ogen)** and **ox(ygen)** and **-yl**. The radical OH.

HYGRO- (before a vowel **hygr-**) Combining form meaning 'moist, humid'. Gk ὑγρο- (hygro-), ὑγρ- (hygr-), fr ὑγρός (hygros), 'wet, moist, humid'.

HYGROSCOPIC adj Compounded by **hygro-** and Gk σκοπείν (scopein), 'to look at, examine'. A term applied to substances which take up moisture from the air.

HYPER- Prefix of Gk origin, generally equivalent to *super-* and *over-*. It usually denotes excess; the highest in a series of compounds. Hyper- is the opposite of **hypo-**. Gk ὑπέρ (hyper), 'over, above, beyond', cognate with OI *upari*, L *super*.

HYPO- (before a vowel **HYP-**) Prefix of Gk origin, generally equivalent to *sub-* and *under-*. It indicates the lowest of a series of compounds. Gk ὑπό (hypo), 'under, below, from below'.

HYPOTHESIS n Mod L fr Gk ὑπόθεσις (hypothesis) 'foundation, supposition', lit 'a placing under', fr ὑπό (hypo), 'under' and τιθέναι (tithenai), 'to put, place'. A provisional supposition which accounts for known facts and serves as a starting point for further investigation by which it may be supported or disproved.

HYPSOCHROME n Gk ὕψος (hypsos), 'height' and χρῶμα (chroma) 'colour'. A group which when introduced into a compound causes its absorption band to shift towards higher frequencies.

HYSTERESIS n Gk ὑστέρησις (hysteresis), 'shortcoming, deficiency', fr ὑστερεῖν (hysterein) 'to be behind, to lag'. The lagging of the dependent variable behind the independent variable. Occurs in magnetisation, adsorption, elastic strain, etc.

I

I Chemical symbol for **iodine** (q.v.), element, at. no. 53.

IATRO- Combining form meaning medical, from Gk ἰατρός (iatros); 'physician'.

IATROCHEMISTRY n From **iatro** (q.v.) and **chemistry**. Chemistry applied to medicine.

-IC Suffix denoting (1) 'pertaining to, of the nature of', as in *catalytic*; (2) a higher valence of the element indicated by the adjective, than that implied by the suffix **-ous**, e.g. *ferric* and *ferrous*. Gk -ικος (-ikos), L *-icus*.

ICE n ME *is*, OE *īs*, related to similar forms in several Teutonic languages.

ICOSA- or **ICOSI-** (before a vowel **ICOS-**) Combining form meaning 'twenty'. Gk εἴκοσι (eicosi), 'twenty'.

ICOSAHEDRON From **icosa-** (q.v.), 'twenty', and Gk ἕδρα (hedra), 'seat, side, face'. A polyhedron with twenty faces.

ICOSANE n From **icos(a)-** (q.v.) 'twenty', and suffix **-ane** denoting a saturated hydrocarbon. The compound $C_{20}H_{22}$.

-ICS Suffix used to form the names of arts and sciences, in imitation of Gk -ικα (-ica), neuter plural of the adj suff -ικος (-icos) which is used to form names of arts and sciences in Greek.

-IDE Suffix used (i) for inorganic compounds containing two elements; (ii) certain organic compounds, mainly anhydrides and lactides; and (iii) some minerals. Back formation from **(ox)ide**.

IGNITE v L *ignitus*, pp of *ignire*, 'to set on fire'. To subject to the action of fire or intense heat ('igniting' a precipitate); to heat to the point of combustion or chemical change.

ILMENITE n Named by Kupfer in 1827 from the *Ilmen* mountains in the Urals. A mineral of composition $FeTiO_3$, the main source of metallic titanium and titanium dioxide.

IMBIBITION n F *imbiber*, fr L *imbibere*, 'to drink in', fr *in-*, 'in' and *bibere*, 'to drink'. A term used in colloid chemistry to denote the absorption of certain liquids by a **xerogel** (q.v.), which leads to the swelling of the gel.

IMIDE A compound of the bivalent radical NH with a bivalent acid radical. Coined through alteration of **amide**, a compound of the univalent radical NH_2.

IMINE n Coined through alteration of **amine**. The name given to the group :NH when the nitrogen atom is part of a ring system or is joined by both valencies to a single carbon atom.

In Chemical symbol for **indium** (q.v.), element at. no. 49.

IN- See list of Latin prefixes (Appendix 3).

-IN Suffix used in a variety of ways in chemical nomenclature, in particular for (1) organic bases, (2) amino-acids, (3) proteins.

INDIGO n F fr Du *indigo*, borrowed from Sp *indico*, *indigo*, fr L *Indicum*, 'indigo', fr Gk ἰνδικόν (indicon), 'indigo', (short for Ἰνδικὸν φάρμακον (Indikon pharmacon), lit 'Indian dye'). Indigo was formerly obtained from the plant *Indigofera tinctoria* which was extensively cultivated in India.

INDIUM n L *indicum*, 'indigo'. The rare metal, at. no. 49, discovered by Reich and Richter in 1863. Named on account of the indigo lines in its emission spectrum. Symbol **In**.

INDOLE n Coined from **ind(igo)** and *ol(eum)*. Lit 'the oil from indigo'. Indole C_8H_7N can be obtained by reduction of isatin, which can in turn be obtained by oxidation of indigo.

-INE Suffix used in a variety of ways in chemical nomenclature, in particular for (1) organic bases, (2) amino-acids, (3) halogen elements.

INERT adj L *iners*, gen *inertis*, 'unskilled, idle, inactive', from L *in*, 'not' and *ars*, 'skill'. Applied primarily to the 'inert gases'.

INFRA Combining form meaning 'under, below'. L *infra*, 'below' (adv), 'under' (prep).

INFRA-RED adj From **infra** (q.v.) and *red*. Pertaining to the part of the spectrum 'below', i.e. of lower frequency than, the red end of the visible spectrum.

71

INHIBIT v From L *inhibitus*, pp of *inhibere*, 'to curb, restrain'. Applied especially to the slowing down of a chemical reaction through the action of a particular substance or substances.

INHOMOGENEOUS adj From L *in*, 'not', and **homogeneous**.

INO- (before a vowel **IN-**) Combining form meaning 'fibrous, sinewy'. Gk ἴς (is), gen ἰνός (inos), 'muscle, fibre'.

INORGANIC adj From L *in*, 'not', and **organic** (q.v.). The chemistry of all elements other than carbon, in contrast to **organic chemistry** (q.v.).

INOSILICATE n From **ino-** (q.v.) and **silicate**. A chain-structured silicate (formed by sharing of oxygens between adjacent tetrahedra). Examples are the pyroxenes and amphiboles.

INOSITOL n Coined fr **in(o)** (q.v.), denoting 'muscle', with **-ose**, for a saccharide; **-ite** (q.v.), and **-ol** (q.v.) for alcohol. A cyclic polyalcohol $C_6H_{12}O_6$ which can be obtained from the cardiac muscular tissue of the ox.

INSECTICIDE n From L *insectum* 'insect', and *-cida* 'killer' from *caedere* 'to kill'. A substance for killing insects.

INSULIN n From *insula* 'island', and suffix **-in**. Name suggested by E. A. W. Shapley, in reference to extraction of the substance from the *islets of Langerhans* in the pancreas.

INTER- Prefix meaning 'among, between'. (See list of Latin prefixes, Appendix 3). L *inter* 'between, among'. (Contrast **intra**, 'within'.)

INTERCALATION n L *inter*, 'between' and *calendae*, 'the first day of the month; month', lit 'the insertion of a day (or month) into the calendar'. Hence the transferred meaning in which elements or compounds with large lattice spacings accommodate other molecules in the holes in the lattice.

INTERMEDIATE adj and n From L *inter-*, 'between' and *medius* 'middle'. Lying in between.

INTERMOLECULAR adj From **inter-**, 'between' and molecular (see **molecule**). Between molecules, as in 'intermolecular forces'.

INTERSTICE n From **inter-** (q.v.) 'between' and the stem of L *stare*, 'to stand'. Lit 'the space between'. A relatively narrow space between the parts of a body, applied particularly to the space between contiguous ions in a crystal lattice.

INTERSTITIAL adj From **interstice**, see prec wd.

INTRA- Prefix meaning 'within, inside'. L *intra-*, 'within, on the inside'. (Contrast **inter**, 'between'.)

INTRAMOLECULAR adj From **intra-** (q.v.), 'within' and molecular (see **molecule**). Within a molecule. (Contrast **intermolecular**.)

INVERSION n From **invert** (q.v.).

INVERT v L *invertere*, 'to turn upside down, reverse'. Applied in particular to the conversion of

72

sucrose into a mixture of glucose and fructose, thereby changing the optical rotation from dextro- to laevo-, i.e. inverting it.

INVERT SUGAR n The mixture of glucose and fructose produced by the hydrolysis of sucrose. See **invert**.

INVERTASE n From **invert** (q.v.) and suff **-ase** denoting an enzyme. The enzyme which converts sucrose into a mixture of fructose and glucose, thereby changing the optical rotation from dextro- to laevo-, i.e. inverting it.

IODATE n From **iod(ine)** and **-ate**.

IODIC adj From **iod(ine)** and **-ic**.

IODIDE n From **iod(ine)** and **-ide**.

IODINE n Gk ἰοειδής (ioeides), 'violet coloured', (fr ἴον (ion), 'violet', -οειδής (-oeides), 'like, resembling', and name ending **-ine**, by analogy with chlorine. Name coined by Davy for the element discovered by Courtois in 1811 in reference to its violet vapour and its resemblance to chlorine. The grey-black solid halogen, element at. no. 53; symbol **I**.

IODO- (before a vowel **IOD-**) Denotes **iodine** (q.v.).

IODOFORM n The **iodine** analogue of **chloroform** (q.v.).

ION n Gk ἰόν (ion), neuter pres p of ἰέναι (ienai), 'to go'. An atom or group of atoms carrying an electric charge. Introduced by M. Faraday, in response to a suggestion by Whewell in 1834; during electrolysis ions move through a solution towards one or other of the electrodes.

IONIC adj From **ion**. (See prec wd).

IONISE v From **ion** (q.v.) and **-ise**, 'to make into'. To convert into an ion or ions.

IR Contraction for **infra-red**.

Ir Symbol for **iridium** (q.v.), element at. no. 77.

IRIDESCENT adj L *iris* 'rainbow', and suff **-escent** (q.v.). Showing rainbow-like colours.

IRIDIUM n Gk ἴρις (iris), gen ἴριδος (iridos), 'rainbow'. White metal, resembling platinum, named by its discoverer Smithson Tennant in 1804 in allusion to the variety of colours of its salts. At. no. 77; symbol **Ir**.

IRIS n L *iris*, fr Gk ἴρις (iris), 'rainbow'. (i) The rainbow, (ii) the coloured portion of the eye, (iii) part of a camera.

IRON n OE *iren* (related to Ger *Eisen*) a word of ancient origin. The abundant and useful metal, element at. no. 26, symbol **Fe** (from L *ferrum*, 'iron').

IRONE n From **ir(is)** and suff **-one**, denoting a ketone. A ketone of the formula $C_{13}H_{20}O$ which is found in the roots of the common iris (*Iris florentina*).

IRRADIATE v L *ir-*, assimilated form before initial r, of *in-*, combining form meaning 'in, inside', and *radiare* 'to shine, beam'. To expose to radiation, visible or otherwise.

-ISE or **-IZE** Suffix forming verbs

to signify (i) 'to make into', as in *crystallise*, or (ii) 'to treat in a certain way'. L *-izare*, fr Gk *-ίζειν* (izein) 'to act in a certain way'.

ISO- (before a vowel **IS-**) Combining form meaning 'equal'. Gk *ἰσο-, ἰσ-* (iso-, is-) from *ἴσος* (isos), 'equal'.

ISOBAR(E) n From **iso-** (q.v.) and Gk *βαρός* (baros), 'weight'. One of two or more elements of the same atomic weight but with different atomic numbers and therefore different properties.

ISOCHORE n From **iso-** (q.v.) and Gk *χῶρος* (choros), 'empty space, room'. Lit 'equal-volume'. The relation between the temperature variation of the equilibrium constant and the enthalpy of a chemical reaction, called by van't Hoff the reaction isochore, because it was originally derived for a constant volume system.

ISOMER n A compound **isomeric** (q.v.) with one or more compounds.

ISOMERIC adj From **iso-**, and Gk *μέρος* (meros), 'part' after the German *isomerisch*, coined by Berzelius. Refers to compounds possessing the same percentage composition and the same molecular weight, but differing in at least one of their physical or chemical properties. Each is an **isomer**.

ISOMORPHOUS adj From **iso-**, Gk *μορφή* (morphe), 'form, shape', and suff **-ous**. Crystallising in the same, or closely related, form. (Introduced by Mitscherlich around 1820, who supposed that equal numbers of atoms combined in the same manner produce the same crystalline form.)

ISOPENTANE n From **iso-** (q.v.) and **pentane** (q.v.). An isomer of pentane, $(CH_3)_2CHCH_2CH_3$.

ISOPHTHALIC adj From **iso-** (q.v.) and **phthalic** (q.v.). Isophthalic acid (also called metaphthalic acid) is an isomer of phthalic acid, $C_6H_4-(COOH)_2$.

ISOPRENE n Coined by C. Greville Williams in 1860 for a reason not stated. The unsaturated hydrocarbon C_5H_8.

ISOPROPYL adj From **iso-** (q.v.) and **propyl** (q.v.). The radical $(CH_3)_2-CH$, isomeric with the propyl radical $CH_3CH_2CH_2$.

ISOSTERE n From **iso-** (q.v.), 'equal', and **stere(o)** (q.v.). Usually encountered as adsorption isostere, the relationship (commonly expressed as a graph) between equilibrium pressure and temperature for a fixed amount of adsorbate.

ISOSTERIC adj See **isostere**. Usually encountered in 'isosteric heat (enthalpy) of adsorption', the differential heat (enthalpy) of adsorption for a given amount adsorbed; this is often calculated from isosteres at neighbouring temperatures.

ISOTHERM n From **iso-** (q.v.) and Gk *θέρμη* (therme), 'heat'. A relationship which is valid at constant temp, e.g. reaction isotherm (van't Hoff 1886); adsorption isotherm.

ISOTHERMAL adj See prec wd. At constant temperature.

ISOTONIC adj Formed from Gk ἰσότονος (isotonos), 'of level pitch', from ἴσος (isos), 'equal' and τόνος (tonos) 'pitch of the voice, tone'. Having equal osmotic pressures.

ISOTOPE n From **iso-** (q.v.) and Gk τόπος (topos), 'place'. One of two or more forms of the same element having the same atomic number but different atomic weights. Introduced by F. Soddy in 1813 in allusion to the fact that isotopes occupy the same position in the Periodic Table of the elements.

ISOTROPIC adj From **iso-** (q.v.) and Gk τρόπος (tropos), 'turn'. Possessing the same physical properties in all directions, in contrast to **anisotropic**.

ITACONIC ACID n Formed fr **aconitic acid** (q.v.) by transposition of letters, since it can be made from aconitic acid (by loss of carbon dioxide). Methylene succinic acid $C_5H_6O_4$.

-ITE Suffix denoting origin or relationship. L *-ita*, *-ites*, from Gk *-ίτης* (ites) 'pertaining to'. Denotes (i) salts of -ous acids; (ii) certain cyclic alcohols, e.g. inosite, quercite; (iii) mineral names.

-IUM Suffix denoting a metallic element, extended by analogy to radicals actually or supposedly resembling a metallic element; but see **helium**.

J

J Symbol for **joule**, unit of energy in the SI system.

JADE n From Sp *piedra de ijada*, 'stone for curing pains in the side.' A pale green ornamental stone at one time believed to be a remedy for colic.

JADEITE n From **jade** (q.v.) and min suff **-ite**. A metasilicate NaAl-$(SiO_3)_2$ which is the chief constituent of jade.

JARGON n OF *jargon*, 'chatter, warbling of birds'. Unintelligible talk. Applied contemptuously to the language of scholars, the terminology of a science, etc.

JASPER n Gk ἴασπις (iaspis), 'jasper', a word of Semitic origin. A hardened siliceous clay, which is used as a gemstone.

JOULE n Named after the physicist *J. P. Joule*. Unit of energy in the SI system. Symbol **J**.

JOURNAL n From F, lit 'a daily paper'.

K

K Chemical symbol for **potassium**, element at. no. 19. Mod L *kalium*, fr Arabic *qili*, 'charred ashes of the glasswort plant' (used in glass making). Also symbol for **Kelvin** (q.v.) the unit of thermodynamic temperature; symbol for equilibrium constant.

k Symbol for the Boltzmann constant. Symbol for the rate constant in chem kinetics. Symbol for **kilo-** (10^3) in the SI system.

KAINITE n Ger *Kainite*, coined by C. F. Zincken in 1865, fr Gk καινός (kainos), 'new', with min suff **-ite**. A mineral of composition $KCl.MgCl_2.6H_2O$.

KAOLINITE n Named in 1867, with min suff **-ite**, fr **kaolin**, a corruption of the Chinese *kauling*, meaning 'high ridge', the name of the hill in China where it is found.

KATHAROMETER n From Gk καθαρός (katharos), 'clean, pure' and **meter** (q.v.). Lit 'purity-meter'; but in fact an instrument for measuring the concentration of a particular gas, present in trace amounts in a stream of carrier gas. Used particularly in gas chromatography.

KELVIN Unit of thermodynamic temperature. After Lord Kelvin.

KERATIN n Gk κέρας (keras), 'horn'. The protein or group of proteins found in skin, hair, nails, feathers, etc.

KETONE n Ger *Keton*, coined by Gmelin in 1848 fr Ger *Aketon*, fr F *acetone*, which was in turn coined fr L *acetum*, 'vinegar', with Gk suff -ωνη (-one), a female patronymic used in chem to denote a weaker derivative.

KIESELGUHR n From Ger *Kiesel*, 'flint' (cf. 'chesil'), and *Guhr*, 'earthy deposit'. A mineral form of silica composed of very fine, hollow particles and used as an adsorbent.

KIESERITE n Ger *Kieserit*, named after *D. G. Kieser*. Hydrous magnesium sulphate, $MgSO_4.6H_2O$.

KILO- Prefix denoting 'a thousand'. Gk χίλιοι (chilioi) 'a thousand'. Symbol (SI system), **k**.

KINETIC adj Gk κινητικός (kinetikos) 'for putting in motion', cf. κινεῖν (kinein), 'to move'. Pertaining to motion, as in 'kinetic energy'.

KINETICS n See prec wd. The study of the rate of chemical reactions.

KNALLGAS n From Ger *knallen*, 'to detonate' and *Gas*, 'gas'. An explosive mixture of hydrogen and oxygen.

Kr Chemical symbol for **krypton**, element at. no. 36.

KRYPTON n Gk κρυπτός (kryptos), 'hidden, concealed'. Name coined by Ramsay and Travers in 1898, for the noble gas, at. no. 36, discovered by them in the air. Symbol **Kr**.

L

L Symbol for the Avogadro constant, after *J. Loschmidt*, who calculated the number of molecules in 1 cm³ of gas.

l Symbol for length.

La Chemical symbol for **lanthanum** (q.v.), element at. no. 57.

LABILE adj ME *labyl*, fr L *labilis*, 'prone to slip'. Unstable.

LABORATORY n Med L *laboratorium*, 'workshop', fr L *laborare* 'to work'.

LACTAM n From **lact(ic)** and **am(ine)**. Lactams are cyclic compounds formed by the elimination of water between the carboxyl group and the amino groups of an amino-acid, and are named by analogy with **lactones** (q.v.).

LACTIC adj L *lac* (gen *lactis*) 'milk'. As in 'lactic acid', $CH_3CH(OH)COOH$ which occurs in sour milk.

LACTIDE n From **lact(ic)** and suff **-ide**. When lactic acid, an α-hydroxyacid, is heated, the OH and COOH groups of one molecule interact respectively with the COOH and OH group of a second molecule, with the elimination of two molecules of water, to give a lactide. The name has been extended to the analagous products from α-hydroxy acids in general.

LACTONE n From **lact(ic)** and

-one. An internal ester formed from a hydroxycarboxylic acid having the OH and COOH groups in appropriate positions. The name coined by Fittig in 1879, reflects the analogy with lactide. (See prec word.) Lactic acid itself does not form a lactone.

LACTOSE n From L *lac*, 'milk' and suff **-ose** denoting a sugar. Milk sugar, $C_{12}H_{22}O_{11}$.

LAEVO- Combining form meaning 'left, to the left', fr L *laevus*, 'left', cognate with Gk λαιός (laios). Applied to the direction of optical rotation as in laevorotatory, now superseded by (−).

LAEVULOSE n Coined by Berthelot fr L *laevus*, 'left', dim suff **-ule** and suff **-ose** denoting a sugar. Fructose or fruit sugar; so called because it is laevorotatory.

LAMBENT adj From L *lambere* 'to lick', cognate with Gk λάπτειν (laptein) 'to lick'. Original meaning 'licking'; now (of a flame) 'playing lightly on a surface without burning it'.

LAMELLA (pl **LAMELLAE**) n L *lamella*, 'a small thin plate', dim of *lamina* (q.v.). A thin plate or sheet.

LAMINA (pl **LAMINAE**) n L *lamina* 'a thin piece of metal or wood; leaf; layer'. A thin plate or scale.

LANOLIN n Coined by Liebrich in 1886 from L *lan(a)*, 'wool'; *ol(eum)*, 'oil, fat'; and chem suff **-in** (q.v.). Wool fat (the palmitate and stearate esters of cholesterol).

LANOSTEROL n From L *lana*

77

(see prec wd), and **sterol** (q.v.). A crystalline alcohol, $C_{30}H_{49}OH$, of the sterol group, which occurs in wool wax.

LANTHANUM n Gk λανθάνειν (lanthanein) 'to escape notice, to lie unseen'; so called because it remained unnoticed in ceria until its discovery by Mosander in 1839. Metallic element, at. no. 57, the first element in the lanthanum series of elements (57–71 inclusive). Symbol **La**.

LANTHANIDE n Any of one of the series of elements with atomic members 57 to 71 inclusive, of which **lanthanum** is the first member.

LAPIDARY n L *lapidarius*, 'pertaining to stone', fr *lapis*, (gen *lapidis*), 'stone', cognate with Gk λέπας (lepas) 'a bare rock, crag'. Pertaining to stones.

LAPIS LAZULI n From L *lapis* 'stone', and gen of ML *lazulum*, fr Arab *lāzaward*, 'blue colour', Pers *lajward*, the word derived from *Lajward*, a place in Turkestan described by Marco Polo. A deep blue semi-precious stone. The powder was used as **ultramarine** (q.v.).

LASER n Coined from initials of Light Amplification (by) Stimulated Emission (of) Radiation.

LATEX n L *latex*, 'liquid, fluid'. A milky juice which exudes from the bark of certain trees, notably *Hevea brasiliensis*, when punctured, and which congeals on exposure to the air (cf. 'rubber latex').

LATTICE n ME *latis*, fr OF *lattis*, fr *latte*, 'lath' which is of Teutonic origin. A structure made of laths fastened together leaving spaces in between. Hence, in crystallography, the regular three dimensional array of atoms or ions in a crystal.

LAURIC ACID n From L *lauris*, 'laurel, bay tree'. The fatty acid $C_{11}H_{23}COOH$, so named because of its occurrence (as glycerides) in the berries of *laurus nobilis*, one of the *laurel* family of plants.

LAURYL adj See prec entry. The radical $C_{12}H_{25}$.

LAW n OE from Scand *lagu*, 'law', cf. ON *log*, 'law', prop plural of *lag*, something laid down or fixed. Hence the extension to a law of nature, or expressible by the statement that a particular phenomenon occurs if certain conditions be present.

LAWRENCIUM n Named after *E. O. Lawrence*, inventor of the cyclotron, and one-time Director of the Berkeley Radiation Laboratory, California, where the element was discovered by Ghiorso, Sükkeland, Larsh and Latimer in 1961. Transuranium element, at. no. 103, symbol **Lw**.

LEAD n OE *lead* (related to Dutch *'lood'*, and Ger *'Lot'*). The metallic element, at. no. 82; Symbol **Pb** (fr L *plumbum*, 'lead').

LECITHIN n Coined by Gobley fr Gk λέκιθος (lecithos), 'yolk of an egg', and suff **-in**. General name for a group of phosphatides, which are an essential constituent of cells, the lecithin from egg yolk being one of the best known.

LEPIDINE n Gk λεπίς (lepis), 'rind, scale'. 4-methylquinoline.

Name coined by Williams (1856) without explanation, but possibly because the compound is related to quinine which is obtained from the bark of the cinchona tree.

LEPIDO- Combining form meaning 'rind, scale, flake'. Gk λεπίς (lepis), gen λεπίδος (lepidos), 'scale, flake'.

LEPIDOLITE n lepido- (q.v.) and min suff -ite. Named by Klaproth in 1794, from the fact that it occurs in scaly masses, rarely in massive crystals. A lithium potassium aluminosilicate with a micaceous appearance.

LEPTON n Gk λεπτόν (lepton) neuter of λεπτός (leptos) 'thin, small, weak'. Any one of a group of elementary particles, including electrons, which are distinguished by relatively weak physical interactions.

LEUCINE n Coined by H. Braconnot in 1820 fr leuco- (q.v.) and chem suff -ine. An amino-acid (α-aminoisocaproic acid), found as a constituent of pancreatic juice and various tissues and organs, named in allusion to its white appearance.

LEUCO- Combining form meaning 'white'. Gk λευκός (leucos), 'white'.

Li Chemical symbol for **lithium** (q.v.), element of at. no. 3.

LIBRATE v L libratus, pp of librare, 'to balance', from libra, 'balance'. Hence 'to oscillate like the arms of a balance'.

LIBRATION n See prec wd. Motion of an oscillating kind.

LIGAND n L ligare, 'to bind, tie'. An oppositely charged ion or neutral molecule, bound to a central metal atom or ion in a coordination compound.

LIGNITE n From L lignum 'wood', and min suff -ite. A low rank, or 'woody', coal.

LIGNOCERIC ACID n From L lignum, 'wood', and L cera, Gk κηρός (ceros), 'wax'. The fatty acid $C_{23}H_{47}COOH$ found both free and combined in many oils, fats and waxes, both animal and vegetable.

LIME n L limus, 'mud', (cognate with Gk λίμνη (limne), 'marsh'. Calcium oxide.

LIMONENE n From Mod L limonum, 'lemon', and suff -ene (q.v.), denoting unsaturation. A terpene, $C_{10}H_{16}$, of lemon-like odour, which like other terpenes is unsaturated.

LIMONITE n Coined by Hausmann (1813) fr Gk λειμών (leimon), 'meadow', related to λίμνη (limne), 'marsh, pool', and min suff -ite. Originally referred to bog iron ore, a brownish yellow deposit common in marshes; now a particular hydrous ferric oxide ($2Fe_2O_3 . 3H_2O$).

LINOLEIC ACID n From L linum, 'flax', and ol(eum), 'oil', with chemical suff -ic. An unsaturated acid, $C_{17}H_{31}COOH$, found as glycerides in linseed, the seed of flax.

LINOLENIC ACID n See prec entry. The insertion of -(e)n(e) denotes that the acid is more unsaturated than linoleic acid. Also occurs as glycerides in linseed oil. Formula $C_{17}H_{29}COOH$.

LIPID n F *lipide*, coined fr Gk λίπος (lipos) 'fat', and suff -id(e). The general name for fats and waxes.

LIQUID adj L *liquidus*, 'flowing, liquid'.

-LITE Combining form meaning 'stone'. F *-lite*, a variant of *-lithe* from Gk λίθος (lithos) 'stone'.

LITHARGE n L *lithargyrus*, fr Gk λιθάργυρος (lithargyros) 'litharge'. Lit 'stone silver', fr λίθος (lithos), 'stone', and ἄργυρος (argyros), 'silver'. (See **litho**, Appendix 6 and **argentum**.) A form of lead oxide which forms on the surface of molten lead.

LITHIUM n Gk λίθος (lithos), 'stone'. Named by its discoverer Arfedson, working in Berzelius' laboratory in 1818, under the impression that the element occurred only in minerals. The alkali metal, at. no. 3, symbol **Li**.

LITHOSPHERE n From Gk λίθος (lithos) 'stone', and σφαῖρα (sphaira) 'ball, sphere'. The solid part of the earth.

LITMUS n ON *litmose*, lit 'lichen for dyeing'. Blue colouring matter obtained from certain lichens, and used as an indicator.

LITRE n F fr Gk λίτρα (litra), which is cognate with L *libra*, both signifying a pound (which consisted of 12 ounces). The L *libra* was also a measure of capacity. A unit of volume in the metric, but not in the SI system.

LOGARITHM n From Mod L *logarithmus*, coined by J. Napier in 1614 from Gk λόγος (logos) 'word, ratio', and ἀριθμός (arithmos) 'number'. Lit meaning is probably 'ratio-number'.

-LOGY Combining form denoting 'discourse, doctrine, science'. Gk λόγος (logos), 'one who speaks (in a certain manner)' or 'one who speaks of (a certain subject)'. Used in names of sciences or departments of subjects.

LUBRICANT n L *lubricans*, pres part of *lubricare* 'to make smooth'. A substance which reduces friction, i.e. renders motion smoother.

LUMINESCENT adj L *lumen*, gen *luminis*, 'light'. Emitting light otherwise than by incandescence.

LUMEN n L *lumen*, 'light'. The unit of luminous flux in the SI system.

LUNAR CAUSTIC n The white crystalline mass formed when silver nitrate is fused and allowed to solidify. So called by the alchemists who termed metallic silver Luna and represented it by the symbol for the moon, possibly because of the pale silvery colour of the moon. Formerly widely used in medicine on account of its mildly caustic properties.

LUTECIUM or **LUTETIUM** n L *Lutetia Parisiorum*, the ancient name of Paris, the native city of Urbain who discovered the element in 1907. Element, at. no. 71, the last member of the lanthanum series. Symbol **Lu**.

Lu Chemical symbol for **lutecium** (see prec wd), element at. no. 71.

Lw Chemical symbol for **lawrencium** (q.v.), element at. no. 103.

LYOPHILIC adj Gk λύσις (lysis), 'a loosening, dissolution', and φίλος (philos) 'loving'. Lit 'solvent-loving'. Term used in colloid chemistry to denote a disperse phase with a high affinity for the dispersion medium, e.g. gelatin in water.

LYOPHOBIC adj Gk λύσις (lysis), φοβία (phobia) 'fear of'. Lit 'solvent-fearing'. Term used in colloid chemistry to denote a disperse phase with a low affinity for the dispersion medium, e.g. gold sol particles in water.

LYSERGIC ACID n Name coined by Jacobs and Craig in 1934: all the *ergot* alkaloids can be hydrolysed to lysergic acid ($C_{16}H_{16}N_2O_2$). The diethylamide of lysergic acid is 'LSD'.

-LYSIS Combining form meaning 'loosening, dissolving, dissolution' as in hydrolysis, pyrolysis. From Gk λύσις (lysis), 'a loosening, dissolution', fr λύειν (lyein) 'to loosen'.

LYXOSE n From *xylose* by transposition of letters (metathesis). A pentose sugar isomeric with **xylose** (q.v.).

M

M Symbol for the prefix **mega** signifying 10^6 in the SI system of units. From Gk μέγας (megas),

'great, large', cognate with L *magnus*, 'great, large'.

m Symbol for **metre** (q.v.), also for **milli-** (q.v.) (10^{-3}) in the SI system.

m- Symbol for **meta-** in organic chemical nomenclature.

MACERATE L *maceratus*, pp of *macerare*, 'to make soft or tender'. To soften by soaking in a liquid.

MACRO- (Before a vowel **MACR-**) Combining form meaning 'long, large'. Gk μακρός (macros), 'long, large'.

MACROMOLECULE n From **macro-** (q.v.) and **molecule** (q.v.). A molecule having a large molecular weight, in excess of 10 000, say.

MACROPORE n From **macro-** (q.v.) and **pore** (q.v.). Applied to pores exceeding ~500 Å (50 nm) in width (cf. Information Bulletin No. 3 of International Union of Pure and Applied Chemistry, Jan. 1970).

MAGMA n Gk μάγμα (magma), 'ointment', from the root of μάσσειν (massein) 'to knead'. Hence a pasty mass, (esp in geology and pharmacy).

MAGNALIUM n Coined from **magnesium** and **aluminium.** An alloy of magnesium and aluminium.

MAGNESIUM n Gk μαγνησία (magnesia), abbreviation of ἡ (he), 'the', Μαγνησία (Magnesia), 'Magnesian', λίθος (lithos), 'stone', i.e. the stone from Magnesia. The name magnesium was at one time used for manganese, but in 1812 Davy adopted the name 'magnesium' for the metal he had newly isolated (in preference to the name 'magnium', which he had first employed). The light metal, element at. no. 12; symbol **Mg**.

MAGNET n L *magnes*, gen *magnetes*, 'lodestone (loadstone)' fr Gk Μαγνῆτις λίθος (Magnetis lithos), 'stone from Magnesia', the name given to a stone with magnetic properties found near each of two places called Magnesia, one in Turkey, one in Northern Greece.

MAGNETON n From **magnet** (q.v.) and **-on**, (q.v.) suff used to denote an elementary particle. Unit of magnetic moment (quantum theory).

MALACHITE n From Gk μαλάχη (malache), 'mallow', and suffix **-ite** denoting a mineral. A green mineral of copper (basic copper carbonate) resembling the leaf of the mallow in colour.

MALACHITE GREEN n A green dyestuff of the triphenyl methane series, resembling the leaf of the mallow in colour (see prec wd).

MALEIC adj See next wd. Maleic acid is an unsaturated acid COOHCH:CHCOOH, which can be prepared from **malic acid** (q.v.). The e denotes the double bond.

MALIC adj From F *malique*, fr L *malum*, 'apple'. Malic acid, $COOHCH_2CHOHCOOH$, occurs in many fruits, esp unripe apples.

MALONIC adj See prec wd. Malonic acid is the dibasic acid $CH_2(COOH)_2$, which can be prepared by oxidation of **malic acid** (q.v.).

MALT n Traceable to the Indo-European word *mel* (hypothetical form), 'to grind', whence Gk μύλη (myle) 'mill', L *molere* 'to grind', cf. *meal* 'edible grain'. Malt is the grain of barley, which has been caused to sprout by being soaked in water and then kept in a moist atmosphere at a suitable temperature.

MALTASE n From **malt** (q.v.) and suff **-ase** denoting an enzyme. The enzyme which brings about the conversion of maltose into glucose.

MALTOSE n From **malt** (q.v.) and suff **-ose** denoting a sugar. A disaccharide $C_{12}H_{22}O_{11}$ made by the action of malt on starch.

MANGANESE n F *manganese*, fr Ital *manganese*, a corrupted form of L *magnesia*, fr Gk Μαγνησία (magnesia) abbreviation of 'the Magnesian stone' (see **magnesium**). The metallic element, at. no. 25, symbol **Mn**.

MANNITOL n From Gk μάννα (manna), gum of several shrubs, esp tamarisk, and **-ite**, and **-ol** denoting an alcohol. Mannitol is found in manna, the dried sap of the manna ash, found chiefly in Calabria and Sicily.

MANNOSE n From Gk μάννα (manna) and **-ose** (q.v.). A monosaccharide obtainable by oxidising **mannitol** (q.v.).

MANOMETER n From F *manomètre*, 'an instrument for measuring that which is thin' coined by Varignon (1654–1722), from Gk μανός (manos), 'porous, scanty' and μέτρον (metron), 'measure'. An instrument for measuring the pressure of gases and vapours.

MARBLE n L *marmor*, fr Gk μάρμαρος (marmaros) 'white glistening stone'. A mineral form of calcium carbonate.

MARGARIC adj Gk μαργαρίτης (margarites), 'a pearl'. Margaric acid was the name given by Chevreul 1813 to a solid fat he obtained from lard, because both it and several of its compounds had a pearly appearance. Actually Chevreul's acid was a mixture; the name today denotes the fatty acid $C_{16}H_{33}COOH$ which does not occur naturally.

MARGARITE n From L *margarita*, fr Gk μαργαρίτης (margarites) 'pearl'. A hydrous silicate of calcium and aluminium with a lustre which is vitreous to pearly. A calcium mica $CaAl_2(OH)_2Si_2Al_2O_{10}$ with a vitreous to pearly lustre.

MASS n From L *massa* 'that which adheres together like dough, lump', fr Gk μᾶζα (maza) 'barley cake, lump' (cf. **macerate**). In science the term signifies 'quantity of matter'.

MASSICOT n From F *massicot*, fr Ital *marzacatto*, 'potters glaze'. Yellow oxide of lead, PbO.

MASURIUM n After *Masuria*, a district in East Prussia, place of origin of the ore in which the element was believed to have been discovered. The name masurium was given to the element of at. no. 43 thought, in error, to have been discovered in 1925 by Noddack, Tacke and Berg. The claim of discovery was later withdrawn. Element 43 was first made, artificially, by Perrier and Segre in 1937 (see **technetium**).

MATERIAL n or adj ME fr Late L *materialis*, 'of matter', fr L *materia*, 'wood, stuff, matter'. See **matter**.

MATTER n ME, *matere*, fr L *materia* 'wood, stuff, matter'. Physical substance in general, as distinct from qualities or energy.

MAXIMUM n L, neuter of *maximus*, 'greatest', from the same source as *magnus*, 'great'. The greatest value of a variable.

Md Chemical symbol for **mendelevium** (q.v.), element at. no. 101.

MECONIC ACID n From Gk μήκων (mekon) 'poppy'. A heterocyclic compound $C_7H_4O_7$ which can be extracted from opium by boiling with water.

MEGA- Prefix denoting 10^6 in the SI system. Gk μέγας (megas), neut μέγα (mega), 'great, large, mighty'.

MELANO- (before a vowel **MELAN-**) Combining form meaning 'black'. Gk μέλανο- (melano-) fr μέλας (melas), gen μέλανος (melanos), 'black'.

MELI- or **MELLI-** Combining form meaning 'honey'. L *mel*, gen *mellis*, 'honey'.

MELISSIC ACID n See **melli-**. The acid $C_{29}H_{59}COOH$, which occurs in beeswax.

MELISSYL ALCOHOL See **melli-**. The alcohol $C_{30}H_{61}OH$ which occurs as palmitate and as melissate in beeswax.

MEMBRANE n L *membrana*, 'fine skin', fr L *membrum* 'limb, member, part'. Original meaning of *membrana* was 'that which covers the members of the body'. A thin, pliable, sheet-like tissue.

MENDELEVIUM n After *Dmitri Ivanovich Mendeleeff* (1834–1907), the Russian chemist, in honour of his work on the periodic classification of the elements. Transuranium element, at. no. 101, made in 1955 by Ghiorso, Harvey, Choppin, Thompson and Seaborg. Symbol **Md**.

MENISCUS n Mod L from Gk μηνίσκος (meniscos), 'a crescent-shaped body', dim of μήνη (mene) 'moon'. In chemistry and physics, the crescent-shaped upper surface of a liquid column caused by capillarity.

MENTHOL n Coined by Oppenheim in 1861 from L *mentha* 'the mint-plant', and suffix **-ol** (q.v.) denoting an alcohol. Menthol $C_{10}H_{19}OH$, a member of the terpene group of compounds, occurs in oil of peppermint.

-MER Suffix denoting 'part', fr Gk μέρος (meros), 'part, share'.

MERCAPTAN n Coined by Zeise in 1834 from ML (*corpus*) *mer-* (*curium*) *captans* 'a substance striving to seize mercury'. (L *captans* is pres part of *captare* 'to try to seize'.) The mercaptans readily react with mercuric oxide to form crystalline compounds, e.g. $(C_2H_5S)_2Hg$.

MERCURY n Named after *Mercury*, the messenger of the gods in Roman mythology; probably on account of its great mobility at room temperature. Element at. no. 80, chemical symbol **Hg**.

MESITYL adj From *mesit*, the name proposed by Reichenbach 1833 for crude acetone obtained by the distillation of wood, fr Gk μεσίτης (mesites), 'the mediator' on account of its great solvent power.

MESITYLENE n From **mesityl** and suff **-ene** denoting unsaturation. The compound 1,3,5-trimethyl benzene, obtainable from acetone by heating with conc. sulphuric acid.

MESO- (before a vowel **MES-**) Combining form meaning 'middle, intermediate'. Gk μεσο- (meso-), fr μέσος (mesos), 'middle'.

MESOMORPHIC adj From **meso-** (q.v.), 'intermediate', and Gk μορφή (morphe), 'form, shape'. A state intermediate between the liquid and the solid, in that it flows, like a liquid, but has some degree of molecular orientation as in a crystalline solid (earlier termed *liquid crystalline*).

MESON n From **meso-** (q.v.) and **-on** (q.v.). One of a series of particles having mass intermediate between that of an electron and that of a proton.

MESOPORE n From **meso-**, 'intermediate', and **pore** (q.v.). A pore intermediate in width between a micropore (width up to $\sim 20\,\text{Å}$ (2 nm)) and a macropore (width above $\sim 1000\,\text{Å}$ (100 nm) IUPAC Information Bulletin No. 3, 1970).

META- Gk μετα- (meta-), 'changed (in form, etc.), next to, between'; as in metaldehyde (a polymer of aldehyde); metaphosphoric acid, meta disubstituted benzene.

METABOLISM n From Gk μεταβολή (metabole), 'change', and *-ism*. Denotes all the chemical changes that occur in a living animal or plant.

METAL n L *metallum*, 'mine, mineral, metal', fr Gk μέταλλον (metallon), 'mine, quarry'.

METALLIFEROUS adj From L *metallum* (q.v.) and the stem of *fero*, 'to bear, carry'. Containing metal(s), as in 'metalliferous ores'.

METAMERISM n From **meta** (q.v.), Gk μέρος (meros), 'part', and suff *-ism*. A term, now rarely used, for the isomerism of the amines; also for functional group isomerism due to differences in the type of compound, e.g. CH_3COCH_3 and CH_3CH_2CHO.

METASTABLE adj From **meta-** and **stable** (q.v.). Term used to describe an equilibrium which is not the most stable equilibrium at the temperature in question, i.e. is a modified equilibrium. Also applied to a substance, e.g. 'a metastable form'.

METATHESIS n Late L fr Gk μετάθεσις (metathesis) 'change of position, transposition', fr μετατιθέναι (metatithenai), 'to transpose', fr μετα- (meta-), 'a change of place or condition', and τιθέναι (tithenai), 'to place, set'. The interchange of atoms or groups between two molecules, the structures of which are not otherwise affected. (Also the interchange between the positions of letters in a word.)

METER n See **metre**.

-METER Combining form denoting a measuring instrument. From L *metrum*, from Gk μέτρον (metron), 'measure'.

METHANE n From **meth(yl)** and suffix **-ane**, denoting a hydrocarbon of the paraffin series. The compound CH_4.

METHYL n From F *méthyle*, back formation from *méthylène*, (see next wd). Coined by Berzelius to denote the radical CH_3.

METHYLENE n From F *méthylène*, coined by Dumas and Péligot in 1835, fr Gk μέθυ (methy), 'wine', and ὕλη (hyle), 'wood', since methyl alcohol can be made by distillation of wood. The radical CH_2.

METRE n fr L *metrum*, fr Gk μέτρον (metron), 'measure'. The unit of length in the SI system (also **meter**).

METRIC adj Pertaining to the **metre** (q.v.), or to the system of weights and measures of which the metre is the unit.

Mg Chemical symbol for **magnesium** (q.v.), element at. no. 12.

MICA n From L *mica*, 'crumb', cognate with Gk μικρός (micros), 'small'. See **micro-**. The mineral mica usually occurs in very small platelets. The development of the English word mica was perhaps influenced by a confusion with L *micare* 'to glisten'.

MICACEOUS adj From **mic(a)-** and suffix **-aceous**. Pertaining to or resembling **mica**.

MICRO- Combining form denoting (i) 10^{-6} in the SI system, (ii) very

85

small. Gk μικρο- (micro-), fr μικρός (micros), 'small'.

MICROANALYSIS n From **micro-**, 'very small', and **analysis**. The chemical analysis of very small quantities.

MICROCOSMIC SALT n F *microcosme*, fr Med L, ult fr Gk μικρός (micros) 'small', κοσμός (cosmos), 'world'. Archaic name of sodium ammonium hydrogen phosphate, presumably denoting the origin in man, microcosm being a term used for man regarded as the epitome of the universe. A crystalline solid obtainable by evaporation of urine.

MICROMETRE n From **micro-** (q.v.), 'one-millionth', and **metre**. One millionth part of a metre.

MICRON n One millionth part of a metre. See **micrometre**.

MICROPORE n From **micro-** (q.v.), 'very small' and **pore**. A very narrow pore, of width less than ~(2 nm) 20 Å. (See IUPAC Information Bulletin No. 3, 1970.)

MICROTOME n From **micro-** (q.v.), 'very small' and Gk τόμος (tomos), 'cutting', root of τέμνειν (temnein), 'to cut'. An instrument for cutting very thin sections for microscopy.

MILLI- Combining form denoting 'one thousandth' (10^{-3}), in the SI system. L *mille*, 'a thousand'.

MILLIGRAMME n From **milli-** and **gramme**. One thousandth of a gramme (10^{-3} g).

MILLILITRE n From **milli-** and **litre**. One thousandth of a litre (10^{-3} l).

MINERAL n From OF *mineral*, F *minéral*, Med L *minerale*, neuter of *mineralis*, 'mineral'. Any substance which is obtained by mining.

MINIMUM n L neuter of *minimus*, 'smallest, least', related to *minuere* 'to make smaller, reduce'. The lowest amount of, or degree of.

MISCIBLE adj From L *miscere*, 'to mix'. Capable of being mixed, e.g. 'miscible with water'.

Mn Chemical symbol for **manganese** (q.v.), element at. no. 25.

Mo Chemical symbol for **molybdenum** (q.v.), element at. no. 42.

MOLE n One gram-molecule, shortened by W. Ostwald (1853–1932). See **molecule**. In the SI system of units, 1 mole is defined as the amount of substance which contains the same number of atoms or ions or molecules as there are atoms in 12 grammes exactly of the pure carbon nucleide ^{12}C.

MOLECULE n F *molécule*, fr Mod L *molecula*, dim of L *moles*, 'mass'. First used in the modern sense by A. Avogadro in 1811. Adj is molecular.

MOLYBDENUM n Mod L *molybdaena* fr Gk μόλυβδος (molybdos) 'lead'. The strange origin of this name is the result of the confusion in ancient times between the various soft black minerals which could be used to make a black streak on a writing surface. Before the 16th century the name molybdenum was applied indiscriminately to the minerals we now know as molybdenite (MoS_2), graphite (C), stibnite (Sb_2S_3), and plumbago or galena (PbS). Then, a difference was observed between plumbago, which yielded metallic lead, and others which did not. In 1779 Scheele showed that one mineral gave carbon dioxide, and the name graphite (q.v.) was adopted for this in 1796. Another mineral on treatment with nitric acid gave sulphuric acid and a peculiar white residue to which Scheele gave the name *acidum molybdenae*. (This confusion between soft black minerals has persisted into our own times; thus crucibles made of graphite have been called plumbago crucibles, lead pencils contain no lead, and in medicine the term 'molybdosis' has meant 'lead poisoning'.) The industrially useful metal, at. no. 42; symbol **Mo**.

MONAZITE n From Gk μονά-ζειν (monazein), 'to live alone', fr μόνος (monos), 'alone'. Monazite (monazite sand), a mineral containing the phosphates of the **rare earths** (q.v.). So called probably on account of the supposed rarity of the mineral.

MONO- (before a vowel, **MON-**) Combining form meaning 'one, alone', from Gk μόνος (monos), 'alone'.

MONOCLINIC adj From **mono-** and κλίνειν (clinein), 'to bend, incline'. Lit 'with one of the axial intersections oblique (inclined)'.

MONOLAYER n Coined by Langmuir (1917), fr **mono-**, 'one' and layer. An adsorbed layer one molecule thick on the surface of a liquid or a solid.

MONOMER n From **mono-** and **-mer**. A molecule which can unite with other molecules of the same substance to give a molecule (polymer) of the same percentage composition but with a molecular weight which is an integral multiple of the original.

MONOMOLECULAR adj From **mono-** and molecular (see **molecule**). Involving one molecule only, as in reference to an adsorbed layer only one molecule thick (monolayer).

MONTMORILLONITE n A clay mineral named by Salvetat in 1847 from the locality Montmorillon, Dept. Vienne, France.

-MORPHIC Combining form meaning 'having a (specified) form'. Formed with adjectival suffix **-ic** from Gk μορφή (morphe), 'form, shape'.

MORPHINE n F formed fr Gk Μορφεύς (Morpheus), name of the god of dreams, son of Sleep, and suff **-ine** denoting an alkaloid. An alkaloid of composition $C_{17}H_{19}O_3N$, named from its ability to induce sleep and to alleviate pain.

MORPHOLINE n See prec wd. A heterocyclic compound C_4H_9NO at one time erroneously believed to have the same ring as in morphine but in fact unrelated to it.

MULLITE n Name coined from *Mull*, Scotland, for a mineral aluminium silicate of composition $Al_6Si_2O_{13}$.

MULTI- Combining form meaning 'much, many', fr L *multus*, 'much, many'.

MULTILAYER n From **multi-** and layer. An adsorbed layer more than one molecule thick on the surface of a liquid or a solid (cf. **monolayer**).

MUSCONE n From musk and suff **-one**, denoting a ketone. The ketonic compound cyclo-[CH_2CO-$(CH_2)_{12}CH(CH_3)CH_2$] found in a secretion from the musk deer.

MUSCOVITE n Named by Dana in 1850 from the old name *Muscovy glass*. A mineral of the mica family.

MYOGLOBIN n From Gk $\mu \tilde{v}_S$ (mys), gen $\mu v \acute{o}_S$ (myos) 'muscle', and **(haemo)- globin** (q.v.). A pigment (muscular haemoglobin) which occurs in muscle.

MYOSIN n Coined from Gk $\mu \tilde{v}_S$ (mys), gen $\mu v \acute{o}_S$ (myos), 'muscle', with suffix. **-in** (q.v.). A protein occurring in muscle.

MYRISTIC ACID n From Linnean generic name, *Myristica*, of the nutmeg tree, fr Gk $\mu v \rho \acute{\iota} \zeta \epsilon \iota v$ (myrizein), 'to anoint', fr $\mu \acute{v} \rho o v$ (myron), 'sweet oil, unguent'. A fatty acid, $C_{13}H_{27}COOH$, found as glyceride in nutmeg oil and other vegetable and animal fats.

N

N Chemical symbol for **nitrogen** (q.v.), element at. no. 7. Symbol for **newton** (SI system).

n Symbol for amount of substance (number of moles). Symbol for **nano-** (q.v.) denoting 10^{-9} in the SI system.

n- Contraction for normal as in n-butane.

Na Chemical symbol for **sodium** (q.v.), element at. no. 11.

NANO- Combining form meaning 'dwarf, dwarfish'. Gk $v \tilde{a} v o s$ (nanos), 'a dwarf'. Used in the SI system of units to denote one thousand-millionth (10^{-9}).

NANOGRAMME n From **nano-** (see prec wd) and **gramme** (q.v.). 10^{-9} gramme.

NANOMETRE n From **nano-** (q.v.) and **metre** (q.v.). 10^{-9} metre.

NANOSECOND n From **nano** (q.v.) and **second** (q.v.). 10^{-9} second.

NAPHTHA n L fr Gk $v \acute{a} \phi \theta a$ (naphtha), 'an inflammable rock-oil obtained from Babylonian asphalt', fr Aramaic *naphtá*. Now applied to most of the imflammable oils obtained by dry distillation of coal, shale and petroleum.

NAPHTHALENE n Originally naphthaline. Coined by J. Kidd in 1821 presumably from **naphtha** (see prec wd) and suff **-ine**, since it was obtained by decomposition of coal tar at red heat. The aromatic hydrocarbon $C_{10}H_8$.

NAPHTHOL n Coined by J. Kidd fr **naphth(alene)** and suff **-ol** denoting a phenol. Either of the two monohydroxy derivatives, $C_{10}H_7OH$, of naphthalene.

NAPHTHYL adj From **naphth-** (**alene**) and **-yl**. The radical $C_{10}H_7$.

NARCOTIC adj and n Gk $\nu\alpha\rho\kappa\omega\tau\iota\kappa\acute{o}s$ (narcoticos) 'numbing'. Having the property of deadening pain, also, a substance which has this property.

NASCENT adj L *nascens*, gen *nascentis*, 'being born'. Applied formerly to a substance at its moment of formation in a chemical reaction, e.g. nascent hydrogen, supposedly atomic.

NATROLITE n Named by Klaproth in 1803 from L *natrium*, 'sodium'. The mineral $Na_2Al_2Si_3O_{10}.2H_2O$.

Nb Chemical symbol for **niobium** (q.v.) element at. no. 41.

Nd Chemical symbol for **neodymium** (q.v.), element at. no. 60.

Ne Chemical symbol for **neon** (q.v.) element at. no. 10.

NEGATIVE adj L *negatus*, pp of *negare*, 'to say no, deny'. The opposite of **positive** (q.v.).

NEMATIC adj See **nemato-**. Thread-like. Applied to the nematic phase of the **mesomorphic** state (q.v.) ('liquid crystals').

NEMATO- Before a vowel **nemat-**, combining form meaning thread. Fr Gk $\nu\hat{\eta}\mu\alpha$, gen $\nu\acute{\eta}\mu\alpha\tau os$ (nematos), 'thread', from the stem of $\nu\acute{\epsilon}\epsilon\iota\nu$ (neein), 'to spin'.

NEO- Combining form meaning 'new, recent'. Gk $\nu\epsilon o$- (neo) fr $\nu\acute{\epsilon}os$ (neos), 'new'.

NEODYMIUM n Gk $\nu\acute{\epsilon}os$ (neos), 'new', and **didymium** (q.v.); in 1885, Auer von Welsbach found that the 'element' didymium, which had been discovered together with lanthanum in 1841 was really a mixture of two elements; he named these 'neodymium' and '**praseodymium**' (q.v.) respectively. The lanthanide element, at. no. 60. Symbol **Nd**.

NEON n Gk $\nu\acute{\epsilon}os$ (neos), 'new', and suff **-on** (by analogy with argon). The 'new' element, present in the atmosphere, discovered by Ramsay and Travers in 1898. The noble gas element, at. no. 10. Symbol **Ne**.

NEOPENTANE n From **neo-**, 'new' and **pentane** (q.v.). The hydrocarbon C_5H_{12}, 2,2-dimethylpropane $C(CH_3)_4$ (see **pentane** and **isopentane**).

NEPHELOMETER n From **nephelo-**, combining form meaning 'cloud', fr $\nu\epsilon\phi\epsilon\lambda o$- (nephelo-), fr $\nu\epsilon\phi\acute{\epsilon}\lambda\eta$ (nephele), 'cloud', and **-meter** (q.v.). An instrument for comparison of the turbidity ('cloudiness') of suspensions, used for chemical analysis.

NEPTUNIUM n After Neptune, the planet beyond Uranus; named in allusion to the fact that this element (at. no. 93) is the 'first beyond' uranium (at. no. 92) which was named after Uranus. The first of the transuranium elements, discovered in the products of irradiation of uranium with neutrons by McMillan and Abelson in 1940. At. no. 93; symbol **Np**.

NEUTRAL adj From L *neuter*, 'neither the one thing nor the other'; formed from negative particle *ne-* and *uter* 'either of two'. (i) Neither positively nor negatively charged; (ii) neither acid nor alkaline.

NEUTRALISE From **neutral** and **-ise** (q.v.). To make neutral.

NEUTRON n An electrically neutral particle, discovered by J. Chadwick in 1932. Coined from **neutr(al)** and **-on**, suff denoting an elementary particle.

NEWTON n After *Isaac Newton*. The unit of force in the SI system.

Ni Chemical symbol for **nickel** (q.v.), element at. no. 28.

NICHROME n An alloy of **ni(ckel)** and **chrom(ium)**.

NICKEL n From Swedish '*Kopparnickel*', fr Ger '*Kupfernickel*', i.e. 'coppernickel' or 'bedevilled copper' (cf. 'Old Nick', the devil), the epithet 'nickel' being used derisively on account of the fact that the ore, in spite of its resemblance to copper ore, yielded no copper (cf. the similar derivations of the names 'cobalt' and 'wolfram'). The metal, at. no. 28, named by its discoverer von Cronstedt, in Sweden in 1751. Symbol **Ni**.

NICOTINE n From Mod L (*herba*) *Nicotiana*, 'herb of Nicot' (=tobacco) from Jean Nicot who introduced tobacco into France in 1560. An alkaloid $C_{10}H_{14}N_2$ found in tobacco leaves.

NIOBIUM n After *Niobe*, the daughter of Tantalus in Greek mythology; name given by Rose in 1844 to the element he discovered, in allusion to its very close resemblance to **tantalum** (q.v.). Element at. no. 41; symbol **Nb**.

NITRATE n Salt or ester of **nitric** acid. See also **-ate**.

NITRE n F *nitre*, fr L *nitrum*, fr Gk *νίτρον* (nitron), perhaps of Oriental origin. The meaning of the term in ancient writings was vague and it stood for different substances. From the early 18th century, 'common nitre' came to mean the substance now known as potassium nitrate.

NITRIC adj See prec wd. Pertaining to nitre, hence, e.g. nitric acid.

NITRILE n From **nitr(o)**, and suffix -*ile*, a variant of **-yl**, denoting a radical. Any ester of hydrogen cyanide; general formula RCN (R = alkyl or aryl radical).

NITRO- (Before a vowel **NITR-**) Combining form used in the sense (i) containing nitrogen; (ii) containing the group NO_2.

NITROGEN n From *nitrogène*, fr *nitro-* 'nitre, potassium nitrate', and -*gène*, by analogy with 'hydrogène', fr Gk *γείνομαι* (geinomai), 'I produce'. Lit 'nitrate producer'; coined by Chaptal in 1790 for the constituent of air otherwise referred to as 'azote' or 'phlogisticated air'. The gaseous element of at. no. 7; symbol **N**.

NITROSYL n Possibly fr *nitrous* and **-yl** (q.v.), nitrosyl being regarded as a derivative of nitrous acid HNO_2. The radical NO.

NOMO- Combining form meaning 'law, usage'. Gk *νομο-* (nomo-) fr *νόμος* (nomos) 'law', lit 'anything settled or assigned'.

NOMOGRAM n From **nomo-** (see prec wd) and **-gram** (q.v.). A type of graphical diagram in which the construction is made once and for all and is applicable to any number of special cases.

NON- Negative prefix. F from L *non,* 'not',

NON-STOICHIOMETRIC adj From **non-** (q.v.) and *stoichiometric* (see **stoichiometry**). Non-stoichiometric compounds are solid phases stable over a range of composition, and therefore not of a unique composition corresponding to a simple chemical formula.

NOR- Combining form denoting a parent compound (regarded as the normal form). The prefix is used to denote that the new compound is derived from the parent form by removal of CH_2, the new compound being regarded as the normal of which the higher is the homologue; e.g. adrenaline $C_8H_9O_3NHCH_3$, noradrenaline $C_8H_9O_3NH_2$. (The prefix is not always used consistently, however.)

NOVO- Combining form meaning 'new'. From L *novus* 'new'.

NOVOCAINE n Coined from **novo-** and **(co)caine.** A local anaesthetic, substitute for cocaine.

Np Chemical symbol for **neptunium** (q.v.), element at. no. 93.

NUCLEAR adj Pertaining to the **nucleus** (q.v.).

NUCLEIC acid(s) n **nucleus.** The non-protein portion of nucleoproteins, so-called because they occur in the *nuclei* of all cells (biology).

NUCLEOPHILE n From **nucleus** (q.v.) and Gk φίλος (philos), 'loving'. Lit 'nucleus-loving'. A reagent which is capable of donating an electron pair to an electron-deficient centre, i.e. to an atomic nucleus, in another molecule.

NUCLEUS n L *nuc(u)leus,* 'kernel', dim of *nux,* gen *nucis,* 'nut'. The central core or kernel, especially of the atom; also the benzene nucleus.

NYLON n An invented word, coined as the name of a synthetic polyamide made from adipic acid and hexamethylenediamine.

O Chemical symbol for **oxygen,** element at. no. 8.

o- Symbol for **ortho-.**

OBSIDIAN n L *Obsidianus lapis* 'the stone of Obsidius', an erroneous reading of Pliny for *Obsianus lapis,* 'stone of Obsius'. A very hard vitreous volcanic rock, discovered in Ethiopia by Obsius.

OCCLUDE v L *occludere* 'to shut up, close up', fr *ob-* and *claudere,* to shut, close'.

OCCLUSION n See prec wd. A rather imprecise term signifying the retention of a gas or of solids by a metal, or the adsorption or absorption of an electrolyte by a precipitate.

OCTA- (before a vowel **OCT-**) Combining form meaning 'eight', Gk ὀκτα- (octa-) fr ὀκτώ (octo), 'eight'.

OCTAHEDRON n Gk ὀκτάεδρον (octaedron), eightsided, fr ὀκτω- (octo-), combining form meaning 'eight' and ἕδρα (hedra), 'seat, base'. A solid figure with eight plane faces.

OCTANE n From ὀκτώ (octo), 'eight' and suff **-ane** denoting an alkane (q.v.). The hydrocarbon C_8H_{18}.

OCTENE n From ὀκτώ (octo), 'eight', and suff **-ene** denoting an unsaturated hydrocarbon. The hydrocarbon of the olefine series C_8H_{16}.

OCTYL n From **oct(a)-** and suff **-yl**. The radical C_8H_{17}.

OHM n The unit of electrical resistance in the SI system. Symbol Ω. After *G. S. Ohm* (1787–1854).

-OID Suffix meaning 'like, having the form of'. Gk -ειδης (-eides), 'like, having the form of', fr εἶδος (eidos), 'form, shape'.

OIL n From L *oleum*, fr Gk ἔλαιον (elaion), 'olive oil'. A greasy liquid.

-OL Suffix used to denote (i) an alcohol, whether simple, or complex (as in glycerol); (ii) a phenol, (since phenols, like alcohols, contain an —OH group); (iii) an oily substance, e.g. indol(e), furfurol, pyrrole, etc. In (i) and (ii) **-ol** is probably derived from the termination of alcohol; and in (iii) is connected with L *oleum*, 'oil', being often written 'ole'.

OLEFIANT GAS n F *oléfiant*, oil making, from L *oleum*, 'oil' and *facere*, to make. Ethylene. (Ethylene combines with chlorine to 'make' an oily liquid, ethylene dichloride.)

OLEFINES n Acyclic hydrocarbons of the formula C_nH_{2n}. So named because they are homologues of olefiant gas (ethylene).

OLEIC ACID n From L *oleum*, fr Gk ἔλαιον (elaion), 'olive oil'. Oleic acid $C_{17}H_{33}COOH$ occurs as glyceride in olive oil.

OLEUM n From L *oleum*, 'oil'. Fuming sulphuric acid, called oleum on account of its oily consistency.

OLFACTORY adj Formed with suff -*ory*, fr L *olfactus*, pp of *olfacere*, 'to smell', which is compounded of *olere*, 'to emit a smell' and *facere*, 'to make'. Pertaining to the sense of smell.

OLIGO- (before a vowel **OLIG-**) Combining form meaning 'a few', fr Gk ὀλίγος (oligos) 'few, scanty'.

OLIGOSACCHARIDE n From **oligo-** (q.v.), 'a few' and **saccharide** (q.v.). Saccharides which yield 'a few' (3 to 6) monosaccharides on hydrolysis.

OLIVINE n Named by Werner in 1790 in allusion to its olive colour. The mineral $(Mg, Fe)_2SiO_4$.

-ON Suffix used (i) to denote the noble gases e.g. argon. Gk -ον (-on), neuter ending of adjectives in ος (os); (ii) as termination for silicon and boron, by analogy with carbon, when the ending comes from L *carbo, carbonis*; (iii) to denote elementary particles as in proton and electron, fr the ending of **ion** (q.v.).

-ONE A suffix denoting a compound of the ketone group, e.g. acetone. Gk -ωνη (-one), a feminine patronymic (name derived from a father or ancestor); in chemistry denotes a *weaker derivative*, e.g. acetone as compared with acetic acid.

-ONIUM Suffix denoting a positively charged radical, e.g. *carbonium*, *phosphonium*. Coined by analogy with ammonium.

OPAL n From L *opalus*, from Gk ὀπάλλιος (opallios), from old Indian *úpalah*, 'stone gem'. A gem stone of iridescent colours, a variety of silica.

OPALESCENT adj From **opal** and suff **-escent**. Like opal in appearance; iridescent.

OPIUM n Gk ὄπιον (opion), 'poppy juice'. The solidified juice of the opium poppy.

OPTIC adj F *optique*, fr Med L *opticus*, fr Gk ὀπτικός (opticos), 'pertaining to the eyes or sight'. Pertaining to the eyes or sight.

OPTICAL adj From **optic** (q.v.) and *-al*. Pertaining to the sense of sight; visual, e.g. optical microscope in contrast to the electron microscope.

OPTIMUM n L *optimus*, 'best'. The best, or most favourable, condition.

ORBIT n From L *orbita*, 'wheel track, rut, path, circuit', fr *orbis*, 'wheel, circle'. The curved path of a satellite about its primary. Hence the

use in chemistry to describe the motion of electrons around the nucleus.

ORBITAL n See prec wd. The geometrical figure which describes the most probable location of an electron.

ORE n ME *oor*, *or*, fr OE *ar*, *aer*, 'brass'. A mineral containing a metal in such quantity as to make its extraction profitable.

ORGANIC adj Gk ὄργανον (organon), 'instrument, bodily organ'.

ORGANIC CHEMISTRY n See prec wd. Originally the chemistry of substances occurring in living organisms; later the chemistry of carbon compounds.

ORIENT n F *orient*, fr L *orientis*, gen of *oriens*, 'the rising sun, east'. The east. See foll wd.

ORIENT v See prec wd. Lit 'to place (anything) so as to face the east'. More generally, to adjust or bring into defined relations; also to ascertain one's bearings.

ORIENTATION n From **orient** (q.v.) and *-ation*. Lit 'the action of turning to or facing the east'. The relative position of atoms or radicals in molecules.

ORIFICE n F, fr L *orificium*, 'an opening', lit 'mouth-making', fr *os*, gen *oris* 'mouth', and *-ficere* 'combining form of *facere*, 'to make'.

ORNITHINE n From Gk ὄρνις (ornis), gen ὄρνιθος (ornithos), 'bird', with suff **-ine**. An amino-acid, $C_5H_{12}O_2N_2$, found in the excrement of birds.

ORPIMENT n F *orpiment*, fr L *auripigmentum*, 'gold paint', which is compounded of *aurum*, 'gold' and *pigmentum*, 'colouring matter'. A mineral form of arsenic trisulphide As_2S_3, which is bright yellow.

ORTHO- (before a vowel **ORTH-**) Combining form meaning 'straight, rectangular, regular, constant'. Gk ὀρθο- (ortho-) from ὀρθός (orthos) 'straight, upright, true, exact'.

ORTHOCLASE n Coined by A. Breithaupt in 1823 fr **ortho-** (q.v.) and Gk κλάσις (clasis), 'fracture', fr κλᾶν (clan), 'to break', on account of its two cleavages at right angles to one another. Potash felspar.

ORTHOGONAL adj Formed with suff -*al* from L *orthogonius*, right-angled, fr Gk ὀρθογώνιος (orthogonios), fr ὀρθός (orthos), 'upright', and γωνία (gonia) 'angle, corner'. Right-angled.

Os Chemical symbol for **osmium** (q.v.), element at. no. 76.

OSAZONE n Name coined by E. Fischer (1884) for the compound formed by reaction of a sugar with phenylhydrazine $C_6H_5NHNH_2$. From **ose** (see **-ose**) for sugar, **az**(**o**) from hydrazine and possibly **-one** on account of the formation of osazones from aldehydes and ketones.

OSCILLATE v L *oscillatus*, pp of *oscillare*, 'to swing, sway'. To swing to and fro.

-OSE Suffix denoting a sugar or other carbohydrate, e.g. glucose, cellulose. From the ending of **glucose** (q.v.) which derives fr Gk γλυκύς (glycys), 'sweet'.

OSMIUM n Coined by its discoverer, the English chemist Smithson Tennant (1761–1815) in 1803 fr Gk ὀσμή 'smell, odour'; so called because of the strong odour of its oxide. The chemical element at. no. 76, chemical symbol **Os**.

OSMOSIS n Mod L, formed with suffix -*osis* fr Gk ὠσμός (osmos) 'a thrusting', fr ὠθεῖν (othein) 'to push'. In osmosis the solvent was supposed to 'push through' the semi-permeable membrane into the solution by virtue of its 'osmotic pressure'.

-OUS Suffix meaning that the element has a lower valency than is denoted by **-ic**. OF -*ous*, -*eus* (F -*eux*), fr L -*osus*, meaning 'having, full of'.

OXALATE n Salt of oxalic acid.

OXALIC n F *oxalique*, fr L *oxalis*, Gk ὀξαλίς (oxalis) 'sorrel', coined by Lavoisier in 1787 because it occurs in the plant sorrel. Pertaining to the acid $(COOH)_2$.

OXIDE n F *oxide* (now *oxyde*), ultimately fr Gk ὀξύς (oxys) and εἶδος (eidos) 'species'. Coined by de Morveau 1790. A compound of oxygen with another element.

OXIDISE v From **oxid**(**e**) and **-ise** (q.v.). To cause to combine with oxygen; to convert into an oxide or oxides; more generally to bring about a process in which an atom or ion loses electrons.

OXONIUM n From **oxy**(**gen**) and **-onium** (q.v.) (by analogy with ammonium). A positive ion in which oxygen is the central atom, e.g. H_3O^+ ($H_2O + HBr$ in SO_2 solution give a solution which ionises into H_3O^+ and Br^-).

OXYGEN n F *oxygène*, short for 'principe oxygène', i.e. 'the acidifying principle', fr Gk ὀξύς (oxys), 'sharp, piercing', and γείνομαι (geinomai), 'I beget, produce' hence 'acid-producing'; name coined by Lavoisier in 1777 to replace the confusing variety of names then in use and to denote the belief of that time that oxygen was the essential constituent of acids. The gaseous element, at. no. 8, symbol **O**.

OZO- Combining form meaning 'smell', usually a 'bad smell'. Gk ὄζο- (ozo-), fr ὄζειν (ozein), 'to smell'.

OZOCERITE or **OZOKERITE** From **ozo-** (see prec wd) and Gk κηρός (ceros) 'beeswax', and min suff **-ite**. A wax-like fossil resin of aromatic odour, comprising a mixture of higher hydrocarbons.

OZONE n Coined in 1840 by C. F. Schönbein from Gk ὄζειν (ozein), 'to smell', on account of its strong smell. The oxygen allotrope, O_3.

OZONIDE n From **ozon(e)** and **-ide**. Ozonides are formed by the addition of ozone to various classes of unsaturated organic compounds.

P

P Chemical symbol for **phosphorus** (q.v.), element at. no. 15.

p Symbol for angular quantum number unity (arising indirectly from the 'principal' series in the atomic spectra of the alkali metals). Symbol for **pico-**, prefix denoting 10^{-12} in the SI system.

p- Symbol for **para-**, as in p-dichlorbenzene.

Pa Chemical symbol for **protactinium** (q.v.), element at. no. 91.

PALLADIUM n After *Pallas*, the asteroid discovered just previously (1802) and named after Pallas Athene, the Greek goddess; name coined by Wollaston in 1803 for the silvery-white metal of the platinum group; at. no. 46; symbol **Pd**.

PALMITIC adj F *palmitique*, fr *palmite*, 'palm marrow', fr Sp *palmito*, 'bud shooting from a palm tree'. Palmitic acid $C_{15}H_{31}COOH$ occurs as glycerides in many fats and oils, including palm oil.

PALMITIN (strictly **TRIPALMITIN**) n From *palmitic* acid (see prec wd) and suff **-in**. The triglyceride of palmitic acid.

PAPAVERINE n L *papaver*, 'poppy', and suff **-ine**, denoting an alkaloid. An alkaloid $C_{20}H_{21}NO_4$ obtained from **opium** (q.v.), which is the solidified juice of the opium poppy.

PARA- (before a vowel **PAR-**) Prefix denoting 'alongside, beyond, similar, near, contrary to', from Gk παρά (para). See list of Gk prefixes, Appendix 4.

PARAFFIN n Name coined by von Reichenbach in 1830 for

paraffin(wax) from L *par(um)*, 'too little' and *affinis*, 'connected with, related to', on account of its lack of chemical reactivity. This name for the paraffin series of hydrocarbons, the alkanes, C_nH_{2n+2} was proposed by H. Watts in 1868.

PARAMAGNETIC adj From **para-** and **magnet(ic)** (q.v.). In a magnetic field, the magnetic dipoles of paramagnetic substances line up parallel to the field (contrast diamagnetic).

PASCAL n After *B. Pascal* (1623–1662). The unit of pressure in the SI system. Symbol **Pa**.

Pb Chemical symbol for **lead** (q.v.), fr L *plumbum*, 'lead', element at. no. 82.

Pd Chemical symbol for **palladium** (q.v.), element at. no. 46.

PECTIN n Coined by H. Braconnot (1781–1855) from Gk πηκτός (pectos) 'congealed, curdled'. Pectins are a group of gelatinous substances obtained from the rind of certain fruits, especially apples, and used in jellies.

PENICILLIN n Coined by Sir Alexander Fleming in 1929, from *Penicillium* (see foll wd), and suff **-in**. The well-known antibiotic prepared from the mould *Penicillium*.

PENICILLIUM n Mod L from L *penicillus* 'painter's brush'. The Penicillium mould, which is so called because it develops organs resembling a brush.

PENTA (before a vowel **PENT**) Combining form meaning 'five'. Gk πεντα- (penta-), fr πέντε (pente), 'five'.

PENTANE n From **pent-** and suff **-ane**, denoting a member of the alkane series of hydrocarbons. Any one of the three isomeric alkanes C_5H_{12}.

PENTENE n From **pent-** and suff **-ene** denoting an unsaturated hydrocarbon. Either of the two alkenes C_5H_{10}.

PENTOSE From **pent-** and suff **-ose** denoting a sugar. A saccharide of the formula $C_5H_{10}O_5$, i.e. containing five carbon atoms.

PEPSIN n Coined by T. Schwann in 1835 from Gk πέψις (pepsis), 'cooking, digestion'. An enzyme secreted in the gastric juices.

PEPTIDE n See **polypeptide**.

PEPTISATION n Gk πεπτικός (pepticos), 'able to digest, promoting digestion'. Coined by T. Graham in 1864 by analogy with the 'peptonisation' of proteins by the digestive juices. The spontaneous dispersion of a precipitate or gel by addition of a small amount of a third substance (peptising agent), or by washing with dispersion medium.

PER- Prefix denoting 'through, by means of', also signifies 'intensity' ('very'). From L *per* 'through, beyond' (and when used as a combining form it intensifies the meaning of the word).

PERACID n From **per-** and **acid**. An acid that may be regarded as a derivative of hydrogen peroxide.

PERI- Prefix meaning 'around, enclosing'. Gk περί (peri), 'around, about, beyond' (in combination it has intensifying force, 'very, exceedingly'). Cognate with L *per* (q.v.).

PERIOD n F *période*, fr L *periodus*, fr Gk περίοδος (periodos), 'circuit, cycle, period of time'; lit 'way round' fr περί (see **peri**) and ὁδός (hodos) 'way'. A portion of time occupied by a recurring process.

PERIODIC adj From **period** and **-ic**. Occurring at regular intervals (cf. 'Periodic System' of the elements).

PERIODIC adj From **per-** (q.v.) and **iodic**. Periodic acid, HIO_4, contains more oxygen than **iodic acid** HIO_3.

PERISTALSIS n Mod L fr **peri** and Gk στάλσις (stalsis), 'compression, constriction'. Rhythmic movement of intestines moving contents on.

PERISTALTIC adj See prec wd. Pertaining to **peristalsis**. The term is used in physical chem to describe a type of circulating pump.

PERMEABLE adj From L *permeabilis*, 'permitting passage through', fr *permeare*, 'to pass through, penetrate' (from **per-** (q.v.), *meare*, 'to go, pass', and suff *-ibilis*, denoting 'able to, tending to').

PEROXIDE n Formed from **per-** (q.v.) and **oxide**.

PERSPEX n Coined (1937) fr L *perspicere*, 'to look through'. Perspex is a transparent polymer.

PERSULPHATE n From **per-** (q.v.) and *sulphate*.

PETRO- Combining form meaning 'stone, rock'. Gk πετρο- (petro-), fr πέτρος (petros), 'stone'.

PETROLEUM n Med L fr L *petra*, fr Gk πέτρα (petra), 'rock', and *oleum*, 'oil'.

pH Symbol for the negative exponent of the hydrogen ion concentration $|H^+|$ so that $|H^+| = 10^{-pH}$. The original symbol P_H^+, proposed by Sörensen in 1909, was simplified to pH by W. M. Clark in 1928.

PHARMACEUTICAL adj Pertaining to **pharmacy** (q.v.).

PHARMACO- Combining form meaning 'drug, medicine, poison', Gk φαρμακο- (pharmaco-) fr φάρμακον (pharmacon) 'charm, philter, drug, remedy'.

PHARMACOLOGY n From **pharmaco-** (q.v.) and Gk λόγος (logos) 'rational explanation'. The science of pharmacy.

PHARMACY n F *pharmacie*, fr Gk φαρμακεία (pharmaceia) 'use of drugs'. The art of preparing and dispensing drugs.

PHASE n A singular formed fr *phases* plural of Mod L *phasis*, fr Gk φάσις (phasis) 'appearance (of a star)' fr the stem of φαίνειν (phainein), 'to make appear, show'. In chem, such bodies as differ in thermodynamic state, or composition, are defined as different phases; all bodies which differ only in quantity or shape are regarded as different examples of the same phase. The term was introduced by J. Willard Gibbs in 1876–8.

PHELLO- (before a vowel **PHELL-**) Combining form meaning 'cork, bark'. Gk φελλο-, φελλ- (phello-), fr φελλός (phellos), 'cork tree, cork'.

PHENACETIN n From **phen(ol)**, **acet(ic)** and suff *-in*. Aceto-p-phenetide, which contains a phenol residue and an acetyl group.

PHENACITE n From Gk φέναξ (phenax), gen φένακος (phenacos), 'cheat, imposter', and min suff -ite. A mineral silicate of beryllium, Be_2SiO_4, at one time confused with quartz.

PHENO- (before a vowel **PHEN-**) Combining form meaning 'pertaining to, or derived from, benzene'. F *phéno*, fr Gk φαίνειν (phainein) 'to make appear, to shine or show'. Introduced by Laurent in 1836: 'I give the name phène (from φαίνω (phaino), 'I light') to the fundamental radical of the preceding acids, since benzene is found in illuminating gas'.

PHENOL n Name introduced by Gerhardt in 1843. From **phen(o)-** (q.v.) and **-ol** (q.v.), denoting either an oily substance or an alcoholic group. The compound C_6H_5OH.

PHENOLPHTHALEIN n From **phenol** and **phthal(ic anhydride)**. The compound $C_{20}H_{14}O_4$ which is prepared by heating phenol and phthalic anhydride in presence of sulphuric acid.

PHENYL n Coined from **phen(o)-** (q.v.) and **-yl**. The radical C_6H_5.

PHENYLHYDRAZINE n From **phenyl** and **hydrazine** (q.v.). The compound $C_6H_5NHNH_2$.

PHLOGISTON n From Mod L, fr Gk φλογιστόν neuter of φλογιστός (phlogistos) 'burnt up, inflammable' (fr Gk φλόξ, gen φλογός (phlogos), 'flame'). Term introduced by Stahl (1660–1734). To the alchemists phlogiston was the principle of inflammability, which escaped when substances were set on fire or underwent oxidation.

PHLOGOPITE n From Gk φλογωπός (phlogopos) 'fiery looking', from φλόξ (phlox) 'flame', and ὤψ (ops), gen ὠπός (opos) 'eye, face'. A magnesium mica with a yellow to brownish lustre.

PHLORIDZIN or **PHLORIZIN** n Gk φλόος (phloos), 'bark', and ρίζα (rhiza) 'root'. A glucoside $C_{21}H_{24}O_{10}$ found in the root bark of apple.

PHLOROGLUCINOL n Gk φλόος (phloos), 'bark', and γλυκύς (glycys), 'sweet', with suff-o l denoting a phenol. Symmetrical tri-hydroxybenzene $C_6H_3(OH)_3$, has a sweetish taste; it was first prepared in 1855 from phloretin which is obtained from the bark of fruit trees.

PHORONE n Coined by Gerhardt fr **(cam)phor** with suff **-one** denoting a ketone. Phorone, $C_9H_{14}O$ a ketone compound resembling camphor in smell and other properties, which is derivable from camphor (oxidation of camphor to camphoric acid, and distillation of the calcium salt).

PHOSGENE n Coined by Humphrey Davy in 1812 from Gk φῶς (phos) 'light', and -γενής (-genes) 'born of'. Phosgene, $COCl_2$, is formed by the action of atmospheric oxygen on chloroform in the presence of light.

PHOSPHATE n From **phosph-(orus)** and **-ate**. A salt of phosphoric acid.

PHOSPHINE n From **phosph-(orus)** and suff **-ine**. Phosphorus trihydride, PH_3.

PHOSPHITE n From **phosph-(orus)** and suff **-ite**. A salt of phosphorous acid.

PHOSPHONIUM n Coined from **phosph(orus)** (q.v.) and the ending **-onium**, on the analogy of ammonium. The radical PH_4.

PHOSPHORESCENT n From **phosphor(us)** (q.v.) and **-escent** (q.v.). Having the property of glowing in the dark, like phosphorus, without actual combustion.

PHOSPHORIC ACID n From **phosphor(us)** and suff **-ic**. Any one of the three acids: ortho-(H_3PO_4), pyro-($H_4P_2O_7$) and meta-(HPO_3); more usually the ortho- form.

PHOSPHORUS n Mod L **phosphorus**, fr L *Phosphorus*, 'the morning star', fr Gk φωσφόρος (phosphoros), 'light bringer', fr φῶς (phos), 'light', and -φορος (phoros) 'bearing', an allusion to the property of glowing in the dark. The nonmetallic element, at. no. 15, symbol **P**.

PHOSPHORYL n Coined from **phosphor(us)** and suff **-yl**. The radical PO.

PHOTO- Combining form meaning (i) light, (ii) photographic. Gk φωτο- (photo-) fr the stem of φῶς (phos), gen φωτός (photos), 'light'.

PHOTOCHEMISTRY n From **photo-** (see prec entry) and **chemistry**. The study of chemical reactions brought about by exposure to visible and ultraviolet radiation.

PHOTOGRAPHY n From **photo-** (q.v.) and Gk γράφος (graphos) fr γράφειν (graphein) 'to write'.

PHOTON n From **phot(o)-** (q.v.) and suff **-on**, denoting an elementary particle. A quantum of light energy.

PHTHALIC adj Abbreviation of (*na*)*phthalic*. Phthalic acid C_6H_4-$(COOH)_2$ can be made by the oxidation of naphthalene.

PHYLLO- (before a vowel **PHYLL-**) Combining form meaning 'leaf', Gk φυλλο-, φυλλ- fr φύλλον (phyllon), 'leaf', cognate with L *folium* 'leaf'.

PHYLLOSILICATE n From **phyllo-** (see prec entry) and **silicate**. Any one of a group of silicates, all of which have a platy or flaky habit and one prominent cleavage; SiO_4 tetrahedra are arranged in layers which are interleaved with other sheets, e.g. of gibbsite. Examples: kaolinite, muscovite.

PHYSICS n L *physica*, fr Gk neut pl τὰ φυσικά (ta physica), lit 'natural things' (the collective title of Aristotle's physical treatises), fr adj

99

φυσικός (physicos), 'pertaining to nature, natural'.

PHYTO- (before a vowel **PHYT-**) Combining form meaning 'plant'. Gk φυτο- (phyto-), fr φυτόν (phyton), 'plant'.

PHYTOL n From **phyt(o)-**, 'plant' and suff **-ol**, denoting an alcohol. The alcohol $C_{20}H_{39}OH$, first isolated by R. Willstätter from the chlorophyll of plants.

PICO- Prefix in the SI system of units (q.v.) denoting 10^{-12}. Ital *piccolo*, 'small'. Symbol **p**.

PICOGRAMME n From **pico-** and **gramme**. 10^{-12} gramme.

PICOLINE n From L *pix*, gen *picis*, 'pitch', suff **-ol**, 'oil' and suff **-ine** denoting a base. One of the three methylpyridines $C_5H_4N(CH_3)$, which occur in coal tar and bone oil.

PICRIC adj From Gk πικρός (picros) 'sharp, bitter'. Picric acid, $C_6H_2(NO_2)_3OH$, is intensely bitter.

PIGMENT n L *pingere* 'to paint', *pigmentum*, 'colouring matter, paint'.

PIMELIC adj Gk πιμελής (pimeles) 'fat' (adj), fr πιμελή (pimele), 'fat, lard'. Pimelic acid, $C_5H_{10}(COOH)_2$, formed by the oxidation of fats.

PINACOL n From Gk πίναξ (pinax), 'plate', an allusion to the crystal shape of the substance, and suff **-ol**, denoting an alcohol. The substance $C_6H_{12}(OH)_2$.

PINENE n From L *pinus*, Gk πίτυς (pitus), 'pine', and suff **-ene**

denoting unsaturation. α-Pinene is a terpene $C_{10}H_{16}$ found in most essential oils derived from *Coniferae*.

PIPERIDINE n From **piper(ine)** (see foll wd) and suff **-idine** fr **(pyr)idine**. Hexahydropyridine, $C_5H_{10}NH$, which can be obtained from piperine by boiling with alkali.

PIPERINE n From L *piper*, 'pepper', and suff **-ine** denoting an alkaloid. The alkaloid $C_{17}H_{19}O_3N$ which is present in *piper nigrin* and other peppers.

PIPERONAL n An *al(dehydic)* compound $C_8H_6O_3$ obtainable by oxidation of **piperine** (see prec wd).

PIPETTE n F *pipe*, 'pipe' and dim ending **-ette**.

PITCHBLENDE n Ger *Pechblende*, name coined by Cronstadt in 1758 fr *Pech*, 'pitch' and *blenden* 'to dazzle'. The chief ore of uranium, UO_2, which is black and often lustrous (hence 'blende') and resembles pitch in appearance.

PLAGIOCLASE n Ger *Plagioclas*, coined by Breithaupt in 1847 fr Gk πλάγιος (plagios), 'oblique, slanting', and κλάσις (clasis) 'fracture'. A triclinic felspar which has two prominent cleavage planes, oblique to one another.

PLASMA n Late L from Gk πλάσμα (plasma), 'something moulded', fr πλάσσειν (plassein), 'to mould'.

PLASTER OF PARIS n So called because it was originally obtained from deposits at Montmartre, *Paris*. Calcium sulphate hemihydrate.

PLATINUM n Sp *platina*, dim of *plata*, 'silver', so named on account of the silvery appearance of the metal. The malleable and ductile metal, at. no. 78, symbol **Pt**.

PLUMBAGO n L *plumbum*, 'lead', *plumbago* 'a species of lead ore'. Name earlier given to graphite since it resembles metallic lead in appearance and consistency.

PLUMBIC adj See foll wd. Containing lead, esp in a state of higher valency.

PLUMBOUS adj L *plumbosus*, 'full of lead'. Containing lead, esp in a state of lower valency.

PLUTONIUM n After *Pluto*, the second planet beyond Uranus, in allusion to the position of this element (at. no. 94) as the second beyond uranium (at. no. 92) in the period classification (cf. neptunium). Transatomic element, first made in 1940–1 by Seaborg, McMillan, Kennedy and Wahl by bombardment of uranium with deuterons, and now manufactured in nuclear reactors for use as nuclear fuel for power production and in nuclear weapons. At. no. 94, symbol **Pu**.

Pm Chemical symbol for **Promethium**, element at. no. 61.

Po Chemical symbol for **Polonium** element at. no. 84.

POLAR adj See **pole**. A polar molecule is one in which the 'centres of gravity' of the positive and negative charges do not coincide, so that the opposite charges form the two poles of a dipole.

POLARIMETER n From **polar** and Gk μέτρον (metron) 'measure'. An instrument for measuring the extent of rotation of the plane of polarisation of light.

POLARISABILITY n See **polarisation**. The extent to which a molecule will suffer separation of its positive and negative electrical centres in an electric field of given strength, to give an induced dipole (dipole moment $= \alpha \times$ field strength, where $\alpha =$ polarisability).

POLARISATION n See **polar**. Molecular polarisation: separation of the positive and negative electrical centres which occurs when a molecule is present in an electric field, resulting in the formation of an induced dipole. Optical polarisation: the modification of the vibrations of light in a particular way. Electrolyte polarisation: effects due to overpotential and concentration charges in the neighbourhood of an electrode.

POLE n L *polus*, 'end of an axis' fr Gk πόλος (polos), 'pivot, axis'. The end of an axis. Magnetic poles: each of the two opposite points or regions on the surface of a magnet at which magnetic forces are manifested (1574). Electric poles: each of the two terminal points (positive and negative) of an electric cell or battery (1802).

POLONIUM n From *Polonia*, the Latinised name of Poland, homeland of Marie Curie (née Sklodowska) who, with her husband Pierre Curie, discovered the element in 1898. Radioactive element, at. no. 84, symbol **Po**.

POLY- Combining form meaning 'much, many'. Gk πολυ- (poly), fr πολύς (polys), neuter πολύ (poly), 'much, many', cognate with L *plus*, 'more', *plurimus* 'most'.

POLYACID n From **poly-** (q.v.) and **acid**. An inorganic acid which may be regarded as derived from two or more molecules of simple acid by elimination of water, e.g. $3H_2CrO_4 - 2H_2O = H_2Cr_3O_{10}$. If the simple acids are of more than one kind, the term heteropolyacid (see **heteropoly**) is used.

POLYMER n See foll wd. A large molecule built up by the repetition of small, simple chemical units.

POLYMERIC adj Ger *polymerisch*, coined by J. J. Berzelius in 1830 fr **poly-** (q.v.) and Gk μέρος (meros) 'part', with suff **-ic**. One compound is polymeric in relation to another if both consist of the same elements in the same proportion by weight, but the molecular weight of one is a multiple of the molecular weight of the other.

POLYMORPHIC adj From **poly-** (q.v.) and μορφή (morphe) 'form, shape'. Denotes that a substance can exist in more than one crystalline form.

POLYPEPTIDE n Coined by E. Fischer in 1906 for substances containing two or more amino-acids joined together by the peptide link —CO–NH–, (now called di-, tri- or polypeptides according to the number of amino-acids). 'The name relates to . . . the old word *peptone*', i.e. the product obtained by digestion of proteins by the action of **pepsin** (q.v.).

PORE n L *porus* from Gk πόρος (poros), 'passage, pore'. Related to

πείρειν (peirein), 'to pierce, or run through'. A minute opening through which fluids may pass.

POROSITY n Med L *porositas*, 'porosity'. The quality of being porous. (See **pore**.)

POROUS adj Having many **pores** (q.v.).

PORPHYRIN n Coined fr Gk πορφύρα (porphyra), 'purple', and suff **-in**. Any one of a group of pigments (important in animal and plant respiration) derived from porphin $C_{20}H_{14}N_4$. (Hopper-Seyler in 1871 obtained a purple pigment by the action of conc sulphuric acid on haemoglobin, and called it haematoporphyrin.)

PORTLAND CEMENT n So called by its inventor J. Aspdin, from its resemblance to Portland stone, quarried in *Portland*, England.

POSITIVE adj L *positivus* 'settled by agreement', fr *positus* 'having been placed', pp of *ponere* 'to place'. Has the sense of being definite; greater than zero; the opposite of negative.

POSITRON n Coined from **posi(tive)** and **elec(tron)**. The positive electron.

POTASH n From Dutch *potaschen* (whence Ger *Pottasche*, Danish *potaske*, Swedish *pottaska*). Lit 'ashes in the pot', because it was originally obtained by evaporation in pots of the clear liquid from extraction of the ashes of vegetable matter with water. Potassium carbonate.

POTASSIUM n From *potassa*, the Latinised form of **potash** (see prec wd). Name coined by Davy in 1807 for the metal he isolated from potash by electrolysis. The alkali metal of at. no. 19, symbol **K** (q.v.).

Pr Chemical symbol for **pr(aseodymium)** (q.v.), element at. no. 59.

PRAGMATIC adj L *pragmaticus* 'skilled in the law', fr Gk πραγματικός (pragmaticos) 'active, businesslike, practical'.

PRASEODYMIUM n Gk πράσιος (prasios) 'leek-green', in allusion to the colour of its salts, and **'didymium'** (q.v.). The lanthanide element separated, together with neodymium (q.v.) from the supposed element 'didymium' by Auer von Welsbach in 1885. Element at. no. 59, symbol **Pr**.

PRECIPITATE v L *praecipitatus*, pp of *praecipitare* 'to throw down headlong'. To deposit, or cause to deposit in solid form from solution in a liquid, by chemical action.

PRECIPITATE n See prec wd. The solid which is precipitated.

PROLINE or **PROLIN** n Coined fr **pyrrole** (q.v.) and suff **-ine**. A derivative of pyrrole of the formula $C_5H_9O_2N$.

PROMETHIUM n After *Prometheus*, the mythological inventor of useful arts who brought fire from heaven. The lanthanide element of at. no. 61, formed in uranium fission and identified by Marinsky, Glendenin and Coryell in 1947. Symbol **Pm**.

PROMOTER n L *promotus*, pp of *promovere*, 'to move forward'. In catalysis, a substance which enhances the catalytic activity of another substance.

PROPANE n From **prop(yl)** (q.v.) and suff **-ane**, denoting a hydrocarbon of the alkane series. C_3H_8.

PROPIONIC adj Coined fr the first syllable of Gk πρῶτος (protos), 'first', and πίων (pion) 'fat', and suff **-ic**. Lit 'first fat'. Propionic acid (C_2H_5COOH) was so called since it is the first acid of the series C_nH_{2n+1}-COOH to have salts with a soapy feel.

PROPYL n From **prop(ionic)** and suff **-yl**. The hydrocarbon radical C_3H_7.

PROPYLENE n Coined from **propyl** and suff **-ene** denoting unsaturation. The hydrocarbon C_3H_6.

PROTACTINIUM or **PROTO-ACTINIUM** n Gk πρῶτος (protos), 'first', and **'actinium'** (q.v.), i.e. 'the precursor of actinium', so named by Hahn and Meitner because, by α-particle emission, it decays radioactively to actinium. The radioactive element of at. no. 91, a homologue of tantalum, discovered independently in 1918 by Soddy and Cranston and by Hahn and Meitner, symbol **Pa**.

PROTEIN n Coined by G. J. Mulder in 1838 fr Gk πρωτεῖον (proteion), 'the first place, the chief rank'. Proteins are essential constituents of every living cell.

PROTO- (before a vowel **PROT-**) Combining form meaning 'first'. Gk πρωτο- (proto-), fr πρῶτος (protos), 'first'.

PROTON n Coined by E. Rutherford in 1920, from **prot(o)-** (q.v.) 'first', by analogy with **(electr)on**. The hydrogen nucleus of unit mass and unit positive charge.

PRUSSIAN BLUE n Transl of F *bleu de Preusse*; so called in allusion to its blue colour and its discovery in Berlin, capital of Prussia (by Diesbach, 1704). Prussian blues are derived from ferric ferrocyanide $Fe^{III}[Fe^{II}(CN)_6]$.

PRUSSIC ACID n From **Prussian blue** (q.v.). The old name for hydrocyanic acid, HCN, since it was first obtained, by Scheele in 1782, by heating Prussian blue with sulphuric acid.

PSEUDO- (before a vowel **PSEUD-**) Combining form meaning 'false, feigned, erroneous'. Gk ψευδο-, ψευδ- fr ψεῦδος (pseudos) 'lie, falsehood'.

PSEUDOMORPH n From **pseudo-** (q.v.) and Gk μορφή (morphe), 'shape, form'. (i) A substance which has changed from one crystal lattice to another, whilst preserving the outer form of the crystal; (ii) a crystal having the external crystalline form of a quite different mineral species.

Pt Chemical symbol for **platinum** (q.v.), element at. no. 78.

PTERIN n Ger *Pterin*, coined from **pter-** (see foll entry) and suff **-in**; so-called because it occurs in the pigments of the wings of butterflies.

PTERO- (before a vowel **PTER-**) Combining form meaning 'feather, wing'. Gk πτερο- (ptero-) fr πτερόν (pteron) 'feather, wing'.

Pu Chemical symbol for **plutonium** (q.v.) element at. no. 94.

PURINE n Coined by E. Fischer in 1884 fr L *purum*, 'clean, pure' and *uricum*, 'uric (acid)', with the suff **-ine**. Purine, $C_5H_4N_4$, may be regarded as the parent body of uric acid.

PURPUREAL adj From L *purpureus*, Gk πορφύρα (porphyra), 'purple'.

PURPUREO- Combining form meaning 'purple' (see prec wd).

PYKNOMETER (**PYCNO-METER**) n From Gk πυκνός (pyknos), 'thick, dense', and **meter** (q.v.). An apparatus for the measurement of the density of a liquid or a solid.

PYRAMID n L *pyramis*, gen *pyramidis*, fr Gk πυραμίς (pyramis), gen πυραμίδος (pyramidos). Prob formed through metathesis fr Old Egyptian *pimar*.

PYRAN n See **pyrone**. α-Pyran and γ-pyran are 6-membered heterocyclic ring compounds corresponding to α- and γ-pyrones in which the methylene group CH_2 replaces the CO group. They have not yet been isolated, but they are the parent substances of a large number of compounds.

PYRENE n Formed fr Gk πῦρ (pyr), 'fire', and suff **-ene**. A white hydrocarbon $C_{16}H_{10}$ obtained from the dry distillation of coal.

PYRIDINE n Coined from Gk πῦρ (pyr), 'fire', and suff **-idine**. Pyridine, C_5H_5N, is formed during the destructive distillation of various nitrogenous organic substances.

PYRIDYL n From **pyrid(ine)** (see prec wd) and suff **-yl**. The radical C_5H_4N.

PYRITE n See foll wd. a Mineral iron disulphide, FeS_2.

PYRITES n L from Gk πυρίτης (pyrites) 'of fire', fr πῦρ (pyr), 'fire'. Any of a number of metallic sulphides, which burn when heated in the air.

PYRO- Combining form fr Gk πῦρ (pyr), 'fire'. It implies (i) that the substance is produced by strongly heating a parent substance; (ii) some property exhibited or alteration produced by the action of heat; (iii) a red or fiery colour.

PYROGALLOL n From **pyro-** implying 'heating', **gall(ic acid)**, and **-ol** denoting phenolic groups. A trihydric phenol $C_6H_3(OH)_3$ which can be produced by heating **gallic acid**.

PYROLIGNEOUS adj A hybrid coined fr Gk πῦρ (pyr), 'fire' and L *lignum*, 'wood'. Pyroligneous acid is the aqueous distillate obtained by the dry distillation of wood.

PYROLUSITE n Ger *Pyrolusit*, coined by Haidinger in 1828. Gk πῦρ (pyr) 'fire' and λούειν (loyein) 'to wash' (which is cognate with L *lavare* 'to wash'), with min suff **-ite**. Mineral manganese dioxide, which is used to neutralise the greenish colour of soda glass (due to iron impurity) by its yellowish colour. It is added to the molten glass, so 'washing it by fire'.

PYROLYSE v See foll wd. To decompose by (rather strong) heating.

PYROLYSIS n Compounded of **pyro-** and Gk λύσις (lysis) 'loosening, setting free'. Decomposition by means of heat.

PYROMETER n Coined by van Musschenbroek (1749) fr **pyro-** (q.v.) and Gk μέτρον (metron) 'measure'. An instrument for measuring (high) temperature.

PYRONE (γ-**pyrone**) n Prob fr *pyrocomane* by abbreviation, and suff **-one**, denoting a keto group. The six-membered heterocyclic compound with one O atom in the ring, $C_5H_4O_2$, originally called pyrocomane (H. Ost, 1884).

PYROPHORIC adj From **pyro-** (q.v.) and Gk φόρος (phoros) 'bearing'. Lit 'fire-bearing'. Applied mainly to powders, particularly of metals, which are so finely divided that they catch fire on exposure to the air.

PYROPHOSPHORIC ACID n From **pyro-** (q.v.) denoting treatment by heat, and **phosphoric acid.** The acid $H_4P_2O_7$ which is obtained by heating orthophosphoric acid at 200–300°C.

PYROTECHNIC adj From Gk πῦρ (pyr) 'fire' and τεχνικός (technicos) 'made by art or skill', fr τέχνη (techne) 'art, skill, craft'. Pertaining to fireworks.

PYROXENE n Coined by Haüy in 1796, fr Gk πῦρ (pyr), 'fire' and ξένος (xenos) 'stranger'. Any one of a group of minerals (mainly silicates of Mg, Ca and Fe), given the name pyroxene under the impression that they were only accidentally caught up in the lavas which contain them.

PYRRHOTITE n Named in 1835 by Breithaupt fr Gk πυρρότης

(pyrrotes) 'redness', and min suff -ite. The mineral $FeS(S)_x$, named in allusion to its colour.

PYRROLE n Ger *Pyrrol*, coined by Runge around 1830, fr Gk πυρρός (pyrros) 'reddish', lit 'flame-coloured', (from the fiery red colour which its vapour imparts to pine splinter moistened in HCl) with suff -ol, 'oil'. The heterocyclic compound C_4H_5N found in bone oil obtained by heating bones.

PYRUVIC ACID n A hybrid word fr Gk πῦρ (pyr) 'fire' and L *uva* 'grape'. The compound $CH_3CO-COOH$ made by pyrolysis of tartaric acid which occurs (as potassium salt) in fermented grape juice. Often called 'pyrotartaric acid'.

Q

Q Symbol for amount of heat.

q Symbol for amount of heat.

QUADRI- or **QUADR-** Combining form meaning 'having four, consisting of four'. L prefix *quadri-*, related to *quattuor*, 'four'.

QUADRIVALENT adj Compounded of **quadri**, 'four' (q.v.) and *valent* (see **valence**). Having a valence of four.

QUADROXALATE n From **quadr(i)-** (q.v.) and **oxalate** (q.v.). Archaic name for compounds of the type $KH_3(COO)_4$, in which four

equivalents of acid are united with one equivalent of base; cf. binoxalate in which there are two equivalents of acid, *n* being added for the sake of emphony.

QUALITATIVE adj L *qualitas*, gen *qualitatis*, 'quality'. Pertaining to quality, often in contrast to quantity (see **quantitative**).

QUANTA n Plural of **quantum** (q.v.).

QUANTITATIVE adj L *quantitas*, gen *quantitatis*, 'greatness, amount'. Pertaining to quantity or amount, often in contrast to quality (see **qualitative**).

QUANTUM n L neuter of *quantus*, 'how much, how many'. Coined by M. Planck in 1900 to denote the elementary quantity, or smallest packet, of energy.

QUARK n Coined by American physicist Murray Gell-Mann in 1961 from James Joyce's *Finnegans Wake* ('Three quarks for Musther Mark'). Sub-atomic particles which, according to the theory, are of three kinds.

QUARTZ n Ger *Quartz*, fr *kwardy*, a W Slav dialectal equivalent of Polish *twardy*, 'hard'. A mineral form of silica, SiO_2, which is very hard.

QUASI- Combining form meaning 'as if, as it were'. L *quasi*, 'as if'.

QUATERNARY adj L *quaternarius*, 'consisting of four each, containing four'. Consisting of four elements.

QUERCITIN n See **quercitrin**. The compound of formula $C_{15}H_{10}O_7$, first obtained by hydrolysis of quercitrin, by Bolley in 1841.

QUERCITOL n L *quercus*, 'oak', and suff **-ol** denoting an alcohol. Cyclohexan-pentol, $C_6H_7(OH)_5$, which occurs in acorns.

QUERCITRIN n From L *quercus*, 'oak' and L *citrus*. A glucoside $C_{21}H_{20}O_{11}$ first discovered, in the bark of the *Quercus tinctoria* (dyer's oak) by Chevreul in 1830.

QUICKLIME n From quick and lime, cf. L *calx viva*. The designation 'quick' alludes to the 'living, active' nature of lime, particularly towards water.

QUICKSILVER n Loan translation of L *argentum vivum*, lit 'living silver'. Mercury, 'quick' alluding to the mobile nature of the liquid metal.

QUINIC ACID n From **quin(ine)** and suff **-ic**. The acid $C_6H_7(OH)_4$-COOH which occurs, in combination, in cinchona bark.

QUININE n Sp *quina*, short for *quinquina*, 'cinchona bark', and suff **-ine**. The base $C_{20}H_{24}N_2O_2$ occurs in all varieties of cinchona bark.

QUINOLINE n Coined by Gerhardt in 1842 fr *quin-* (see **quinine**), -ol, (q.v.), 'oil', and suff **-ine**. An oily nitrogeneous base C_9H_7N, obtained by Gerhardt by fusing quinine with caustic potash.

QUINONE n From **quin(ic) acid** (q.v.) and suff **-one** denoting keto groups. The compound $C_6H_4O_2$, which contains two keto groups, was first obtained (by A. Woskresensky) by oxidation of quinic acid with MnO_2 and H_2SO_4.

QUINQUE (before a vowel **QUINQU-**) Combining form meaning 'five'. L *quinque*, 'five'.

QUINTESSENCE n F, fr Med L *quinta essentia*, 'the fifth (and finest) essence or element', loan translation of Gk πεμπτή οὐσία (pempte oysia), 'fifth substance'. (In ancient mediaeval philosophy this was the substance of which the heavenly bodies were made and which was latent in all things.) A highly refined essence or extract.

R

R Symbol for the molar gas constant.

Ra Chemical symbol for **radium**, element at. no. 88.

RACEMIC adj See **racemic acid**. The term has been generalised to denote a mixture in equal proportions of the dextro-(+) and laevo-(−) rotatory forms of any optically active compound.

RACEMIC ACID n L *racemus*, 'a cluster of grapes'. Named coined by Gay-Lussac in 1828 for the optically inactive form of tartaric acid which had been isolated from tartar of grapes; it is a mixture of the dextro-(+) and laevo-(−) rotatory forms of tartaric acid, in equal proportions.

RADICAL n L *radicalis*, fr *radix*, 'root'. A group of elements which can enter into or be expelled from combination, without itself under-going decomposition.

RADIOACTIVE adj See **radium**. Having the property of emitting radiation in the form of α, β and γ rays.

RADIUM n L *radius*, 'ray', in allusion to the property of emitting energy in the form of rays. The radio-active element isolated from pitch-blende by Pierre and Marie Curie in 1898; at. no. 88, symbol **Ra**.

RADON n From **rad(ium)** (q.v.) and **-on** (by analogy with argon, krypton, etc.), so named because it is the noble gas emitted by radium in the course of radioactive decay. Radioactive element, at. no. 86; symbol **Rn**.

RAMAN EFFECT n After *C. V. Raman*. An effect in which the light scattered at right angles to the incident radiation contains frequen-cies differing from that of the original radiation, which are characteristic of the substance being irradiated.

RARE EARTHS n Oxides of the 'rare earth metals', fourteen metallic elements of closely similar properties, Ce, Pr, Nd, Pm, Sm, Eu, Gd, Tb, Dy, Ho, Er, Tm, Yb, Lu, at. no. 58–71, in Group IIIA, of the Periodic Table; so called because all were at one time believed to be of rare occurrence. Now frequently termed lanthanide elements, since they closely resemble lanthanum, at. no. 57.

RATE n L *rata* 'settled', fem pp of *reor, reri*, 'to reckon, suppose, judge'; cf. *pro rata parte*, 'proportionally, according to a fixed amount.' Speed; change per unit time.

RAYON n F *rayon*, 'ray'. Arti-ficial silk, named from its glossy appearance.

Rb Chemical symbol for **rubidium**, (q.v.), element at. no. 37.

Re Chemical symbol for **rhenium**, (q.v.), element at. no. 75.

RE- L prefix meaning '*back, again*', and implying (i) repetition of action, (ii) restoring to original condition, (iii) turning back, (iv) intensification.

REACTION n From **re-** (see prec wd), and *action*, fr L *actio*, gen *actionis*, 'action'. The action of one substance on another, or the result of such action.

REAGENT n From **re-**, (q.v.) and *agent*, fr L *agens*, gen *agentis*, pres p of *agere* 'to do, act'. A substance which acts on another substance.

REALGAR n Named by Wallerius in 1747 fr Arabic *rahj al-ghär*, 'powder of the mine', fr *rahj*, 'pow-der', *al*, 'the' and *ghär*, 'cave, mine'. Mineral As_2S_2.

RECTI- (before a vowel **RECT-**) Combining form meaning 'straight'. L *rectus*, 'straight, right'.

RECTIFY v L *rectus*, 'right' and *facere*, 'to make'. (i) To distil a volatile liquid to free it from im-purities. (ii) To transform alternating current to direct current.

REFRACT v L *refractus*, pp of *refringere*, 'to break up', fr **re-** (q.v.) and *frangere*, 'to break'. To bend a ray, especially of light.

REFRACTORY adj L *refractarius* 'stubborn, obstinate'. Resists the action of heat, is difficult to fuse or work in any way.

REFRIGERANT n L *refrigerans*, gen *refrigerantis*, pres p of *refrigerare* 'to make cold'. A substance used to bring about cooling.

RESEARCH v (also n) From **re-** (q.v.) and *search*. To investigate closely.

RESIDUE n F *residu*, fr L *residuum*, neuter of *residuus*, 'that which is left behind'. The solid (or liquid) left behind after extraction, distillation, etc.

RESIN n F *resine*, fr L *resina*, fr Gk ῥητίνη (rhetine), 'resin of pine'. Originally applied only to vegetable products, now also to any organic or silico-organic material characterised by high molecular weight and a gummy or tacky consistency at certain temperatures.

RESONANCE n F *résonance*, fr L *resonantia*, 'echo'. (i) Absorption by a system of radiation which it is capable of emitting; (ii) a means of visualising electronic distributions in molecules by considering the contributions from certain extreme structures (canonical forms) which do not themselves exist; the actual arrangement is thus a hybrid of the canonical forms.

RESORCINOL n Coined by its discoverers Hlasiwetz and Barth in 1864 fr **res(in)** (q.v.) and *orcin*; the suff **-ol** was added later in allusion to its phenolic nature. (*Orcin-* fr Ital *orcello*, 'archil', a dyestuff derived from lichens.) 3,5-dihydroxy-1-methylbenzene.

RETORT n From **re-** (q.v.) and L *torquere*, 'to twist'. A vessel with a neck which is bent back, or 'twisted'.

REVERBERATE v From L *reverberare*, 'to strike back, to cause to rebound'. To throw back (see foll wd).

REVERBERATORY adj See prec wd. A reverberatory furnace is one in which the flame is forced back on to the substance exposed to it, i.e. on to the bed of the furnace.

Rh Chemical symbol for **rhodium** (q.v.) element at. no. 45.

RHAMNOSE n L *rhamnus*, 'buckthorn'. (-)Rhamnose $C_6H_{12}O_5$ is a sugar which can be extracted from the bark of *Rhamnus frangula* or the berries of *Rhamnus catharticus*.

RHENIUM n L *Rhenus*, 'the river Rhine', name given to element at. no. 75 by its discoverers Noddack, Tacke and Berg in 1925. The rare metal, homologue of manganese and technetium; symbol **Re**.

RHEO- Combining form meaning 'stream, or current'. Gk ῥέος (rheos), 'anything flowing, stream' fr the stem of ῥεῖν (rhein), 'to flow'.

RHEOLOGY n From **rheo-** (q.v.) and **-logy** (q.v.). The science of the deformation and flow of materials.

RHEOSTAT n Coined by Wheatstone (ca. 1843) fr **rheo-** (see prec entry) and στατός (statos), 'placed, standing'. A regulating device (resistor) used so as to obtain a constant current.

RHODIUM n Gk ῥόδον (rhodon), 'rose', so named in allusion to the colour of its salts. The rare metal of the platinum group discovered by Wollaston in 1804. At. no. 45, symbol **Rh**.

RHODOCROSITE n Named by Hausmann in 1813 fr Gk ῥοδόχρως (rhodochros), 'rose coloured', an allusion to its colour. A mineral form of manganese carbonate.

RIBONIC ACID n By rearrangement of some of the letters of *arabonic acid* of which it is a stereoisomer. D-Arabonic acid can be prepared by oxidation of D-arabinose, which in turn can be extracted from gum arabic.

RIBOSE n See **ribonic acid**. A pentose, $C_5H_{10}O_5$, the laevo-(−) form of which can be obtained by reduction of D-ribonic acid.

RICIN n From L *Ricinus*, 'the castor oil plant'. A protein found in the seeds of the castor bean.

RICINOLEIC ACID n fr L *Ricinus*, 'the castor oil plant', and suff **-ol** denoting an oil. An unsaturated acid $C_{18}H_{34}O_3$ occurring as glyceride in castor oil.

Rn Chemical symbol for **radon** (q.v.), element at. no. 86.

ROTATE v L *rotatus*, pp of *rotare*, 'to go round in a circle', fr *rota*, 'wheel'. To turn about an axis (see **rotation**).

ROTATION n From **rotate** with noun ending *-ion*. The action of moving round a centre. Also applied to turning of the plane of polarisation of light by an optically active compound.

ROTATORY adj Formed with suffix *-ory* from **rotate** (q.v.).

Ru Chemical symbol for **ruthenium** (q.v.), element at. no. 44.

RUBIDIUM n L *rubidus*, 'darkest red', so named by its discoverer Bunsen in 1861 in allusion to the two red lines in its emission spectrum. The rare alkali metal of at. no. 37; symbol **Rb**.

RUBY n F *rubis* fr Med L *rubinus*, fr L *ruber, rubeus*, 'red'. A deep-red gem, a form of alumina.

RUTHENIUM n Med L *Ruthenia*, 'Russia', a metal first observed in platinum from Russia by Osann in 1828 but first definitely identified by Claus in 1844 and for which he retained Osann's name 'ruthenium'. The rare metal of the platinum group; at. no. 44; symbol **Ru**.

RUTILE n From Ger '*Rutil*', coined by Werner in 1803 fr L *rutilis*, 'red, shining, glittering' (related to L *ruber*, 'red'), in allusion to its colour. A mineral form of titania, TiO_2.

S

S Chemical symbol for **sulphur**, element at. no. 16. Symbol for entropy. Symbol for (surface) area.

s Symbol for second in the SI system. Symbol for angular momentum of zero (indirectly arising fr the 'sharp' series in the atomic spectra of the alkali metals).

SACCHARIDE n From Med L *saccharum,* Gk σάκχαρον (saccharon), 'sugar'. From OI *sarkara,* 'gravel, grit, sugar'. Saccharides are polyhydric alcohols of sweet taste, and contain one or more five- or six-membered rings each containing an oxygen atom.

SACCHARINE or **SACCHARIN** n Coined by Fahlberg and List fr Med L *saccharum,* 'sugar'. The sweet-tasting compound, $C_7H_5O_3NS$, used as a substitute for sugar.

SALICIN n F **salicine,** coined from L *salix,* gen *salicis,* 'willow' and chem suffix **-ine.** A glucoside $C_{13}H_{18}O_7$ which was first isolated from willow bark.

SALICYL adj See **salicylic acid.** The radical C_6H_4OH.

SALICYLIC ACID n L *salix,* 'willow'. The acid, $C_6H_4(OH)COOH$, which was prepared in 1838 from **salicin** (q.v.) obtained from willow bark.

SALT n OE *sealt,* related to *salt* in several Scand languages, to Ger *Salz,* etc.; L *sal.* Sodium chloride ('common salt'); more generally a substance produced by the interaction of equivalent quantities of an acid and a base.

SAMARIUM n After *samarskite,* a mineral (named after *von Samarski,* a Russian mining official) in which it was found by Lecoq de Boisbaudran in 1879. The lanthanide element of at. no. 62, symbol **Sm.**

SANDWICH n Named after the Fourth Earl of Sandwich (1718–1792), who once spent 24 hours at the gaming table, eating only this kind of food. Used in chemistry to denote a structure in which a layer of species B is bonded between two layers of A.

SANTONIN n L *herba santonica,* 'the herb of Santoni', European wormwood. $C_{15}H_{18}O_3$ an unsaturated ketonic lactone which can be extracted from *santonica.*

SAPONIFICATION n See **saponify.**

SAPONIFY v F fr L *sapo,* gen *saponis,* 'soap', and *-ficere* fr *facere,* 'to make, do'. Originally, to make soap from glycerides by boiling with alkalies; then extended to the hydrolysis of other esters by alkalis even though the product was not a soap. The term *hydrolysis* is now used in preference.

SAPPHIRE n ME fr OF, fr L *sapphirus* fr Gk σάπφειρος (sapphei-ros), fr Hebrew *sappir,* 'sapphire, lapis lazuli' (q.v.). A blue, transparent variety of crystalline corundum Al_2O_3.

SARCOSINE n Ger *Sarkosin,* coined by Liebig in 1847 from Gk σάρξ (sarx) gen σαρκός (sarcos), 'flesh', and suff **-ine.** A crystalline amino acid, methyl glycine CH_3-$HNCH_2COOH$, first obtained (by Liebig) by boiling creatine—which occurs in muscles and nerves—with water.

SATURATED adj From L *saturatus,* pp of *saturare,* 'to fill', which is related to *satis,* 'enough'. (i) Of a solution; being in equilibrium with

Sb

Sepiolite

the solid or liquid solute. (ii) Of a vapour; being in equilibrium with the liquid. (iii) Of a compound; being incapable of undergoing addition reactions.

Sb Chemical symbol for **antimony** (q.v.) element at. no. 51.

Sc Chemical symbol for **scandium** (q.v.) element at. no. 21.

SCALE n ME from L *scalae* (pl), 'ladder, flight of steps, staircase' related to *scandere* 'to climb'. Hence a succession of steps or degrees (cf. a scale in music) and thus a series of graduations.

SCALES n ME fr ON *skal*, 'shell, dish, scale of a balance' related to Ger *Schale*, 'drinking cup, bowl; scale of a balance'. A weighing instrument.

SCANDIUM n Med L *Scandia*, 'Scandinavia'. The element, predicted under the name of ekaboron by Mendeléeff in 1871 and discovered by Nilson in 1879. At. no. 21; symbol **Sc**.

SCHEELITE n Named by von Leonhard in 1821 in honour of *K. W. Scheele*, the Swedish chemist who discovered tungstic acid, with min suff **-ite**. Mineral form of calcium tungstate, $CaWO_4$.

SCORBUTIC adj Mod L *scorbuticus*, fr *scorbutus*, 'scurvy'. Pertaining to scurvy. (See **ascorbic acid** Vitamin C.)

Se Chemical symbol for **selenium** (q.v.), element at. no. 34.

SECOND n F *seconde*, from Med L *secunda*, contraction of L *pars*

minuta secunda, lit. 'the second small part', i.e. the result of the second division of the hour by sixty. The unit of time in the SI system. Symbol **s**.

SEDIMENT n F *sediment*, fr L *sedimentum* 'a settling down, subsidence', fr the stem of *sedere* 'to sit'. Matter composed of particles that fall by gravitation to the bottom of a liquid suspension.

SELENITE n L *selenites*, fr Gk σεληνίτης (selenites), short for σεληνίτης λίθος (selenites lithos), 'moon stone'. A transparent, mineral form of gypsum $CaSO_4$, probably named in allusion to the pale bluish reflection of the mineral. Also any salt of selenious acid.

SELENIUM n Gk σελήνη (selene), 'the moon', so named by its discoverer Berzelius in 1817 on account of its resemblance to tellurium (q.v.) which had been named after the Earth (L *tellus*). At no. 34,. symbol **Se**.

SELININE n L *selinon*, Gk σέλινον (selinon) 'celery'. A sesquiterpene $C_{15}H_{24}$ which occurs in celery oil.

SEMI- combining form meaning 'half'. L *semi* 'half', cognate with Gk ἡμι- (hemi-).

SEMIPERMEABLE adj From semi- (see prec entry) and **permeable** (q.v.). Semipermeable membranes are so called since they permit the passage of some molecules (usually of the solvent) but not of others (usually of the solute).

SEPIOLITE n From Gk σήπια (sepia), 'cuttlefish', and min suff **-ite**.

112

Mineral $H_4Mg_2Si_3O_{10}$, named by Glocker in 1849 from the fact that it is light and porous like cuttlefish bone and floats on water.

SEPTUM n L *saeptum, septum,* 'wall, enclosure'. A (thin) partition. In chem (esp gas chromatography) a membrane of rubber, etc. which can be readily pierced.

SERINE n L *sericum,* 'silk'. An amino-acid $NH_2CH(CH_2OH)COOH$ which occurs in silk gelatin.

SERPENTINE n ME, fr F *serpentin,* 'like a serpent', fr L *serpens,* gen *serpentis,* 'serpent'. A hydrous magnesium silicate (mineral) of a dull green colour with markings resembling those of a serpent's skin.

SESQUI- Combining form meaning 'one half more', fr L *sesqui,* 'and a half more'. In chem means 'one and a half'. Now obsolescent except in sesquiterpenes, formula $C_{15}H_{24}$, as compared with terpenes, $C_{10}H_{16}$.

SI UNITS 'Système Internationale' units. The system based on the six units (symbols in brackets): metre (m); kilogramme (kg); second (s); ampere (A); kelvin (K); candela (cd); mole (mol). (For SI prefixes see Appendix 8.)

Si Chemical symbol for **silicon**, (q.v.) element at. no. 14.

SIDERITE n F *siderose,* name given by Beudant in 1832 and changed to Ger *Siderit* by Haidingen in 1845. Gk σιδηρίτης (siderites), 'of iron', fr σίδηρος (sideros) 'iron'. Native ferrous carbonate.

SILANE n From **sil(icon)** and suff **-ane,** (q.v.) usually denoting a saturated hydrocarbon. Silanes are compounds of the formula Si_nH_{2n+2}, and were so named by A. Stock in 1916 on account of their formal analogy to the alkanes, C_nH_{2n+2}.

SILICA n L *silex,* gen *silicis,* 'flint, pebble stone'. Silicon dioxide (flints are concretions of silica).

SILICATE n From **silica** and suff **-ate.** A salt of silicic acid.

SILICON n L *silex,* gen *silicis,* 'flint, hard stone', and **-on** (by analogy with carbon and boron); name proposed by Thomas Thomson in 1831 for the non-metallic element present in flint and having a resemblance to carbon and boron (Berzelius had called this element silicium or kiesel). The element of at. no. 14; symbol **Si.**

SILICONE n From **silic(on)** and suff **-one,** by analogy with the designation of ketones. Originally applied to compounds of formula R_2SiO (R = alkyl group) analogous to simple ketones; now a generic term to cover all compounds containing Si and organic groups, provided Si is present in sufficient amount to affect the properties appreciably.

SILICOSIS n Medical. Coined from **silic(a)** and **-osis,** a medical suffix of Gk origin. A disease caused by the inhalation of silica (quartz).

SILVER n From OE *silofr,* ultimate origin unknown. The precious metal, element at. no. 47, symbol **Ag** (fr L *argentum,* 'silver').

SINTER v Prob fr OE *sinder,* 'impurity of metal, dross', related to similar words in Germanic languages.

To adhere in a mass as a result of heating below the melting point.

SITO- Combining form meaning 'food'. Gk σιτο- (sito-) fr σῖτος (sitos), 'wheat, corn, meal'.

SITOSTEROL n From **sito-** (see prec wd) and **sterol** (q.v.). The sterol $C_{29}H_{50}O$ found in wheat germ oil.

SKATOL(E) n Coined fr the stem of σκῶρ (skor) gen σκατός (skatos), 'dung'. The compound C_9H_9N, found in faeces and urine. (The suff **-ol** (q.v.) usually denotes an alcohol or an oil; skatole is neither.)

Sm Chemical symbol for **samarium** (q.v.) element at. no. 62.

SMECTITE n Gk σμηκτίς (smectis), 'fuller's earth', fr σμήχειν (smechein), 'to cleanse'. Smectite is a variety of fuller's earth, a clay mineral used for removing oily matter from wool.

Sn L *stannum*, 'tin'. Chemical symbol for **tin**, element at. no. 50.

SODA n Possibly fr L *solida* 'a solid'. Sodium carbonate.

SODALITE n A hybrid coined by T. Thomson in 1811 fr **soda** (q.v.) and Gk λίθος (lithos), 'stone'. A mineral, $3NaAlSiO_4 \cdot NaCl$, of the felspar group named in allusion to its sodium content.

SODIUM n From **soda** (q.v.), so named by Davy in 1807 because it was obtained by electrolysis of caustic soda (sodium hydroxide). The common alkali metal, element at. no. 11; symbol **Na** (fr Mod L *natrium*, fr *natron* 'mineral sodium carbonate').

SOLDER n Med E *soudour*, *soulder*, fr F *souder* 'to solder', fr L *solidare* 'to make solid'. A fusible metallic alloy used for joining metallic surfaces.

SOLID n (also adj) L *solidus*, 'firm, compact'. A body characterised by rigidity.

SOLID SOLUTION n From **solid** and **solution**. A homogeneous mixture of two solids; a molecular, atomic or ionic dispersion of one solid in another.

SOLUTE n L *solutus* pp fr *solvere* 'to loosen'. A dissolved substance.

SOLUTION n L *solutio*, gen *solutionis*, 'a loosening, dissolution'. A molecular, ionic or atomic dispersion of one substance on another.

SOLVATION n From **solv(ent)** and suff **-ation**, denoting process or condition. The attachment of one or more molecules of solvent to an ion in solution.

SOLVE v L *solvere*, 'to loosen, untie, dissolve; to solve'. To dissolve.

SOLVENT adj and n From **solve** (q.v.) and suff **-ent** denoting an agent. A substance having the power of dissolving other substances.

SOOT n OE *sōt*, related to ON *sot* and Old Dutch *soet*, lit 'that which settles'. A deposit consisting of fine particles formed by the combustion of fuels.

SORB v L *sorbere*, 'to suck in, to swallow up'. To take up by adsorption or absorption. (Introduced by J. W. McBain in 1909.)

SORBIC ACID n L *sorbus*, 'service tree, mountain ash'. An unsaturated acid $CH_3CH:CHCH:CHCOOH$ which can be obtained from the berries of mountain ash.

SORBITOL n See prec wd. A hexahydric alcohol, $HOCH_2(CH-OH)_4CH_2OH$, present in small amounts in the berries of mountain ash, and stereo-isomeric with mannitol.

SORBOSE n See prec wd. A ketose sugar formed by the oxidation of sorbitol.

SORPTION n The act of sorbing (see **sorb**).

SPATULA n L *spatula*, dim of *spatha*, 'a broad tool or weapon', fr Gk σπάθη (spathe), 'a broad blade of wood or metal'. A simple flat instrument of metal etc. for stirring mixtures and other purposes.

SPECTRO- Combining form for **spectrum** (q.v.).

SPECTROGRAPH n From **spectro-** (q.v.) and **-graph** (q.v.). An instrument for displaying a spectrum.

SPECTROMETER n A hybrid coined from L *spectrum*; 'appearance, image' and Gk μέτρον (metron), 'measure'. An instrument for measuring the angular deviation of radiation, e.g. by passing through a prism.

SPECTROSCOPE n A hybrid coined from **spectrum** (q.v.) and Gk σκοπεῖν (scopein), 'to look at, examine'. An instrument for forming and analysing spectra.

SPECTROSCOPY n The art of using the **spectroscope** (q.v.).

SPECTRUM n L spectrum *'appearance, image'*, fr *specere*, 'to see, look at'. Originally the coloured band ranging from red to violet, obtained when white light passes through a prism. Now used to include a wide range of electromagnetic radiation.

SPECULUM n From L *specere*, 'to see, look at'. A mirror.

SPHALERITE n Named by Glocker in 1847 from Gk σφαλερός (sphaleros), 'treacherous, deceitful' (since it was easily confused with other minerals, especially galena (PbS)) with min suff **-ite**. An ore of zinc, ZnS. (See **blende**.)

SPHERE n ME *spere*, fr Gk σφαῖρα (sphaira), 'ball, globe, sphere'.

SPHEROID n From **sphere** (q.v.) and adj suff *-oid*, meaning 'like, resembling'. A body resembling, but not identical with, a sphere. Hence adj spheroidal.

SPINEL n Ital *spinella*, dim of *spina*, fr L *spina*, 'thorn, spine'. Etymology uncertain. A group of double oxides with the general formula $R^{II}O. R_2^{III}O_3$.

SPINTHARISCOPE n Coined fr Gk σπινθαρίς (spintharis), 'spark' and σκοπεῖν (skopein) 'to look at, examine'. An instrument for observing the scintillation caused by X-rays striking a fluorescent screen.

SPIRO- Combining form meaning 'spiral, whorled'.

SPONTANEOUS adj Late L *spontaneus*, 'of one's own free will', fr L *sponte*, 'voluntarily'; first used by the English philosopher Thomas Hobbes in 1656. A spontaneous process is one which takes place without the aid of an outside agency.

SQUALENE n L *squalus*, 'a large fish'. A dihydrotriterpene $C_{30}H_{50}$ which occurs in the livers of sharks of the family squalidae.

Sr Chemical symbol for **strontium** (q.v.) element at. no. 38.

STABLE adj L *stabilis*, 'firm, steady, constant', from the stem of *stare*, 'to stand'. Presenting resistance to displacement; not easily decomposing.

STALACTITE n Mod L *stalactites* fr Gk σταλακτός (stalactos) 'dropping, dripping' fr σταλάσσειν (stallassein) 'to let drop, to drop'. An icicle-shaped formation (especially of calcium carbonate) hanging from the roof of a cave and formed by the dropping of waters which have percolated through limestone.

STALAGMITE n Mod L *stalagmites*, fr Gk σταλαγμός (stalagmos), 'a dropping'. A cone-shaped deposit on the floor of a cave formed by the dropping from the roof of some material (especially calcium carbonate) in solution.

STATIC adj Mod L, *staticus* fr Gk στατικός (statikos) 'causing to stand'. Pertaining to bodies at rest, or to forces in equilibrium; opposite to dynamic.

STEARATE n A salt of **stearic acid** (q.v.).

STEARIC ACID n See **stearin**, whence it is obtained by hydrolysis. The fatty acid $C_{17}H_{35}COOH$.

STEARIN n F *stearine*, coined by Chevreul, (ca. 1813) fr Gk στέαρ (stear), 'fat'. A general name for the three glycerides (mono-, di-, and tristearin) formed by the combination of stearic acid and glycerine; mostly applied to tristearin, the chief constituent of tallow and suet.

STEEL n ME *stel*, fr OE *stile*, *stel*, *style*, and similar words in Ger and Scandinavian languages. The original meaning was probably something firm and hard.

STEREO- (before a vowel **STER-**) Combining form meaning 'solid, firm'. Gk στερεο- (stereo-) fr στερεός (stereos), 'firm, hard, solid'. In chem has the sense of 'in space, spatial'.

STEREOISOMERISM n From **stereo-** (q.v.) and isomerism (see **isomeric**). It may be either: (i) geometrical isomerism in which the isomers differ in the spatial arrangement of certain groups; or (ii) optical isomerism, when the arrangement of the atoms is such that a molecule and its mirror image are not superposable.

STERIC adj From **ster(eo)-** (q.v.) and suff **-ic**. Spatial.

STEROID n From **ster(ol)** (q.v.) and suff **-oid**, meaning 'resembling, like'. A generic name for substances having the same, or very similar, ring structure to the sterols.

STEROL n From **ster(eo)-** (q.v.) denoting 'solid, firm' and suff **-ol** denoting an alcohol. Any one of a group of unsaturated alcohols, having a ring structure, which are solid (cf. **cholesterol**).

STIBNITE n Coined by J. D. Dana in 1854 from Beudart's name *stibine*, fr L *stibium*, 'antimony'. A mineral form of antimony trisulphide.

STIGMASTEROL n A sterol (q.v.) which occurs in *Phytostigma venosum*, the Calabor bean. Formula $C_{29}H_{48}O$.

STILBENE n F *stilbène*, coined by Laurent in 1843 fr Gk στίλβειν (stilbein), 'to glitter', from the brilliancy of the crystals, and suff -ene, denoting a double bond. A crystalline hydrocarbon, $C_6H_5CH: CHC_6H_5$.

STOICHIOMETRY n Coined by J. B. Richter in 1792 fr Gk στοιχεῖον (stoicheion), 'element', and -μετρια (-metria) 'process of measuring'. The application of the law of conservation of matter and the laws of combining weights to chem processes. A 'stoichiometric compound' is one in which the ratio of the number of atoms to one another as determined from the atomic weights is a ratio of small whole numbers.

STREPTOMYCIN n From *streptomyces griseus* with suff -in. (Gk στρεπτο- (strepto-) 'easy to bend, pliant', and μύκης (myces) 'fungus'.) A complex organic base, $C_{21}H_{39}O_{12}N_7$, produced by the mould *Streptomyces griseus*, and used as an antibiotic.

STRONTIANITE n Named by Sulzer in 1791 from the locality *Strontian* in Argyllshire, Scotland, where it is found. Natural strontium carbonate.

STRONTIUM n From *Strontian*, Argyllshire, source of the mineral strontianite (strontium carbonate) and, hence, of strontia (strontium oxide) from which Davy in 1808 obtained the metal by electrolysis. The alkaline–earth metal, at. no. 38; symbol **Sr**.

STRYCHNINE n F, fr L *strychnos* 'nightshade', fr Gk στρύχνος (strychnos), 'nightshade', and suff -ine. An alkaloid, $C_{21}H_{22}N_2O_2$, so-called because it was first found (in 1818) in the plant *Strychnos Ignati*.

STYRENE n From *stor(ax)* with suff -ene, denoting a double bond. The hydrocarbon $C_6H_5CH:CH_2$ first produced by pyrolysis of *Storax*, a liquid resin from the *Liquid-amber orientalis tree*.

SUB- Prefix of Latin origin meaning 'under'.

SUBERIC adj F *subérique*, fr L *suber*, gen *suberis*, 'cork'. Pertaining to cork.

SUBERIC ACID n See prec wd. A dibasic acid $HOOC(CH_2)_6COOH$, which can be obtained by the action of nitric acid on cork.

SUBSTANCE n ME fr F, fr L *substantia*, 'that of which a thing consists', fr *substare* 'to stand or be under, to be present'.

SUBSTRATE n ML *substratum*, fr **sub-** (q.v.) and *stratus*, pp of *sternere* 'to strew, spread out'. In chemistry the sense has been broadened to mean 'the substance being acted upon', particularly when a catalyst is present.

SUCCINIC adj F *succinique,* formed with suff *-ique,* fr L *succinum, sucinum,* 'amber'. A dibasic acid $HOOC(CH_2)_2COOH$ which occurs in amber.

SUCROSE n From F *sucre,* 'sugar', with suff -ose denoting a carbohydrate. The disaccharide $C_{12}H_{22}O_{11}$ commonly known as cane sugar.

SUGAR n ME *suger* fr F *sucre,* fr Med L *succarum,* fr Arab *súkkar,* cf. Ital *zucchero,* Sp *azúcar,* Ger *Zucker.* (The initial *a* in *azúcar* is the Arabic article *al.*) The disaccharide sucrose; also the lower **saccharides** (q.v.) in general.

SULPHUR or **SULFUR** (U.S.) n L *sulpur* or *sulfer,* 'sulphur'. The non-metallic element, at. no. 16. Symbol **S**.

SUPER- Prefix meaning 'above, over'. L *super,* 'over, above, on top of; beyond'; cf. Gk ὑπέρ (hyper), 'over, above, beyond'.

SUPERCOOL v A hybrid word, formed fr **super-** (q.v.) and *cool.* When a liquid is cooled below its freezing point, but fails to solidify, it is said to be supercooled, i.e. cooled overmuch.

SUPERHEAT v A hybrid word fr **super-** (q.v.) and *heat.* When a liquid is heated above its boiling point but fails to boil it is said to be superheated, i.e. heated overmuch.

SUPERPHOSPHATE n From **super-** (q.v.) and **phosphate**. A mixture of $Ca(H_2PO_4)_2 . H_2O$ and $CaSO_4 . 2H_2O$ which, being more soluble than rock phosphate $(Ca_3(PO_4)_2)$ is more effective as a fertiliser.

SUPERSATURATED adj From **super-** (q.v.) and **saturated**. Description applied to a solution when it contains more solute than that of a saturated solution at the same temperature, or to a vapour when its vapour pressure is higher than the saturated vapour pressure at that temperature.

SUS- A form of **sub-** used in some Latin compound words in which the second element begins with *s, p* or *t.* (See list of Latin prefixes, Appendix 3).

SUSPEND v F *suspendre,* fr L *suspendere,* 'to hang up', fr **sus-** (q.v.) and *pendere,* 'to hang'. To hold in the form of particles diffused throughout a liquid or gaseous medium.

SUSPENSION n See prec wd. The state of being suspended; the system of suspended particles plus medium.

SYDNONE n Coined fr *Syd(ney)* and suff **-one**. A heterocyclic ring compound, so called because it was discovered (by Earl and Mackney in 1935) in Sydney. It is representative of a group of substances having a high dipole moment and an apparently divalent nitrogen atom.

SYLVITE n Named by Beudant from the chemical *sal digestevus Sylvia* in 1832. Mineral form of potassium chloride.

SYMMETRY n L *symmetria,* fr Gk συμμετρία (symmetria), 'due proportion', fr σύμμετρος (symmetros), 'of like measure'. Exact correspondence in position of the several parts of a body with reference to a dividing line, plane or point.

SYN- Prefix meaning 'with, together with'. Gk σύν (syn) 'with, together'. In geometrical isomerism, it implies that two given groups lie on the same side of the plane of a double bond (cf. **anti-**.)

SYNERESIS n From Gk συναίρεσις (synairesis) 'contraction'. Coined by T. Graham in 1864 to denote the spontaneous contraction of a gel with loss of dispersion medium, usually water.

SYNTHESIS n L fr Gk σύνθεσις (synthesis), fr συν- (syn-) and θέσις (thesis), 'a placing, arranging'. Formation of a compound by combination of its elements or constituents.

SYNTHETIC adj See prec wd. Produced by **synthesis** (often, as opposed to occurring naturally).

T

T Chemical symbol for **tritium** (q.v.); symbol for **thermodynamic temperature**; symbol for **tera-**, prefix denoting 10^{12} in the SI system of units.

t Symbol for time.

t- Symbol for tert-, see **tertiary**.

Ta Chemical symbol for **tantalum**, element at. no. 73.

TALC n F fr Sp *talco*, fr Arab *talaq, talq*. A soft mineral, magnesium silicate.

TANNIC ACID n F fr Mod L *tannum*, (which is of Celtic origin) 'oak tree'. A phenolic compound, $C_{14}H_{10}O_9$, which is found in oak-galls.

TANTALUM n After Tantalus, son of Jupiter, who was condemned to remain thirsty though standing chin-deep in water, for the water receded whenever he stooped to drink it. The element so named by its discoverer Ekeberg in 1802 because of the tantalising difficulties experienced in getting it into solution. The dense metal, of at. no. 73; symbol **Ta**.

TAR n OE *teru* 'pertaining to a tree', related to similar words in other Germanic languages, prob derived ultimately from an IE root meaning 'tree', cf. Gk δρῦς (drys) 'tree, oak'. A viscous liquid obtained by destructive distillation of wood, coal or other organic substances.

TARTAR n Med L *tartarum*, Med Gk τάρταρον (tartaron), perhaps of Arabic origin, name given to the deposit (potassium hydrogen tartrate) adhering to the sides of wine casks. Potassium hydrogen tartrate.

TARTARIC adj Pertaining to tartaric acid.

TARTRATE adj A salt of tartaric acid.

TAURINE n From L *taurus*, 'bull' and suff **-ine**. The substance $NH_2CH_2CH_2SO_3H$ discovered by Gmelin in 1824 in ox-gall, where it occurs as taurocholic acid.

TAUTO- Combining form meaning 'the same', fr Gk ταυτο- (tauto-), contraction of τὸ αὐτό (to auto), 'the same'.

TAUTOMER n From **tauto-** (see prec entry) and **-mer** (q.v.) 'part'. When a compound behaves as if it had two or more different chem structures, each is called a tautomer.

TAUTOMERIC adj Pertaining to, or exhibiting, **tautomerism**.

TAUTOMERISM n See **tautomer**. The property of a compound whereby it behaves as if it had two or more different chem structures.

Tb Chemical symbol for **terbium**, element at. no. 65.

Tc Chemical symbol for **technetium**, element at. no. 43.

Te Chemical symbol for **tellurium**, element at. no. 52.

TECHNETIUM n Gk τεχνητός (technetos) 'artificial', so named because the element is not found in nature and was made artificially, by Perrier and Segrè by the bombardment of molybdenum by deuterons. Element at. no. 43; symbol **Tc**.

TELLURIUM n L *tellus*, gen *telluris*, 'earth', so named by Klaproth in 1799 to signify the contrast between this element and the one he had discovered previously and named

uranium (q.v.). The non-metallic element of at. no. 52; symbol **Te**.

TEMPERATURE n L *temperatura*, 'due measure, proportion', first used in its present sense by G. Galileo around 1590.

TER- Combining form meaning 'three times'. L *ter*, 'three times'.

TERA- Prefix in the SI system of units denoting 10^{12}. Cf. Gk τέρας (teras) 'monster'. Symbol **T**.

TERBIUM n From *Ytterby*, Sweden, from which place was obtained the mineral gadolinite in which Mosander in 1843 discovered the element (cf. **erbium, yttrium**). The rare lanthanide element, at. no. 65; symbol **Tb**.

TEREPHTHALIC adj As in terephthalic acid. Coined fr *tere(bene)* and **phthalic** (q.v.), so-called because it was obtained by oxidising oil of turpentine (L *terebinthus*, fr Gk τερέβινθος (terebinthos) 'the terebinth or turpentine tree'). Benzene-p-dicarboxylic acid $C_6H_4(COOH)_2$.

TERPENE n Shortened from *terpentine* the original form of turpentine (q.v.). Any one of a group of hydrocarbons $C_{10}H_{16}$, notably pinene and limonene.

TERTIARY adj From L *tertius*, 'third', and adj suff *-ary*.

TERYLENE n Coined from **tere(phthalic)** and suff **-ene**. A synthetic fibre made from terephthalic acid.

TETRA- (before a vowel **TETR-**) Combining form meaning 'four'. Gk τετρα- (tetra-), 'four'.

TETRAGONAL adj L *tetragonum,* fr Gk τετράγωνος (tetragonos), 'with four angles', fr **tetra-** (q.v.) and γωνία (gonia), 'angle', with adj suff -*al*. Having four angles.

TETRAHEDRAL adj See next wd.

TETRAHEDRON n L fr Late Gk τετράεδρος (tetraedros), 'having four sides', fr **tetra-** (q.v.) and ἕδρα (hedra) 'seat, base'. A solid figure having four sides.

TETROXALATE n See **quadroxalate.**

TEXTILE adj and n L *textilis,* 'woven', fr *textus,* pp of *texere,* 'to weave, plait'. Woven (adj); woven material (n).

Th Chemical symbol for **thorium,** element at no. 90.

THALLIUM n Gk θαλλός (thallos), 'young shoot, twig', so named by Crookes in 1861 in reference to the bright green line in its spectrum. The rare metal of at. no. 81, symbol **Tl.**

THEINE n From *Thea,* a genus of plants of the tea family, and suff -**ine,** denoting an alkaloid. An alkaloid $C_8H_{10}N_4O_2$, identical with **caffeine** (q.v.) found in tea.

THEOBROMINE n From Mod L *Theobroma,* name coined by Linnaeus in 1760 for a genus of trees of the chocolate family, fr Gk θεο- (theo-), 'god', and βρῶμα (broma) 'food'. Lit 'the food of the gods'. An alkaloid, (cf. suff -**ine**), $C_7H_5N_4O_2$, occurring in cacao beans, from the tree *Theobroma Cacao.*

THEORY n Late L *theoria,* fr Gk θεωρία (theoria), 'a viewing, consideration, contemplation', fr θεωρεῖν (theorein) 'to see, contemplate'. A **hypothesis** (q.v.) that has been confirmed by experiment or observation.

THERMAL adj Gk θερμός (thermos) 'hot'. Pertaining to heat.

THERMAL ANALYSIS See **thermal** and **analysis.** Analysis by the controlled application of heat.

THERMO- (before a vowel **THERM-**) Combining form meaning heat. From Gk θερμός (thermos), 'hot'.

THERMOCHEMISTRY n From **thermo-** (q.v.) and **chemistry.** The quantitative study of the heat evolved or absorbed during chemical reactions.

THERMODYNAMICS n From **thermo-** (q.v.) and **dynamics** (q.v.). Literally the word implies a study of the relationship between heat and mechanical energy, but it has been broadened to embrace the interconversion of the various forms of energy.

THERMOMETER n From **thermo-** (q.v.), and -**meter** (q.v.). An instrument for measuring temperature.

THERMOSTAT n From **thermo-** (q.v.) and Gk στατός (statos) 'placed, standing'. An instrument for maintaining a constant temperature.

THESIS n L fr Gk θέσις (thesis) 'a setting, placing, arranging'.

THIAZOLE n From **thi(o)-** (q.v.) and **-azole** denoting nitrogen. The thiazole ring contains 3 carbon, 1 nitrogen and 1 sulphur atom. Thiazole is C_3H_3NS.

THIO- (before a vowel **THI-**) Combining form denoting 'containing sulphur'. Gk $\theta\epsilon\hat{\iota}o\nu$ (theion), 'brimstone'.

THIOCYANATE n From **thio-** (q.v.) and **cyanate** (q.v.). A salt of thiocyanic acid, HSCN.

THIONIC ACIDS n See **thio-**. The acids $H_2S_nO_6$ ($n = 2$ to 6).

THIONYL adj From **thio-** (q.v.) and suff **-yl**. The radical SO:.

THIXOTROPY n From Gk $\theta\acute{\iota}\xi\iota\varsigma$ (thixis), 'a touch', and $\tau\rho\sigma\pi\acute{\eta}$ (trope), 'a turning, change'. The property of becoming fluid when stroked or shaken, as shown by some gels.

THORIUM n After *Thor*, Scandinavian god of thunder, name coined by Berzelius in 1828. The radioactive metallic element of at. no. 90, symbol **Th**.

THULIUM n From *thulia*, the oxide named by Cleve in 1879 after *Thule*, the ancients' name for 'the most northerly land'. The rare lanthanide element discovered in thulia by Lecoq de Boisbaudran in 1886; at. no. 69; symbol **Tm**.

THYMOL n From L *thymum*, fr Gk $\theta\upsilon\mu\acute{o}\nu$ (thymon), 'thyme', and suff **-ol** denoting a phenolic compound. The aromatic substance $C_{10}H_{13}OH$ which occurs in oil of thyme.

THYROID adj Gk $\theta\upsilon\rho\epsilon\sigma\epsilon\iota\delta\acute{\eta}\varsigma$ (thyreoeides) 'shield-shaped' fr $\theta\upsilon\rho\epsilon\acute{o}\varsigma$ (thyreos), 'oblong shield', and suff **-oid** (q.v.). The largest cartilages of the larynx consist of two plates thought to resemble oblong shields. The thyroid gland is close to the larynx.

THYROXINE n Coined by its discoverer E. C. Kendall in 1915 fr **thyr(oid)** (q.v.), **-ox** and suff **-ine**. The substance, $C_{15}H_{11}O_4NI_4$, which occurs in the thyroid gland.

Ti Chemical symbol for **titanium**, element at. no. 22.

TIN n OE *tin*, ultimate origin unknown. The silver-white metal, familiar in tin plate; element at. no. 50; symbol **Sn** (fr L *stannum*, 'tin').

TINCTURE n L *tinctura*, 'dyeing', from *tinctus*, pp of *tingere*, 'to dye'. A tinge, or tint. In alchemy, a tincture was an active principle contained in, or derivable from, a substance.

TINT n From L *tingere*, 'to dye' (see **tincture**). A colour, usually slight or delicate.

TITANIUM n Gk $T\iota\tau\hat{a}\nu\epsilon\varsigma$ (Titanes) 'the Titans', the giants of Greek mythology, name coined by Klaproth in 1794 (in reference to his discovery a few years before of uranium and to the fact that in Greek mythology the Titans were the sons of Uranus). The light metallic element of at. no. 22; symbol **Ti**.

TITRATE v From F *titrer* 'to give a title to; to determine the strength', fr *titre* 'title, right; fineness of alloyed gold or silver'. To determine the 'strength' (concentration) of a solution by finding the volume of a standard solution which reacts with it.

TITRATION n See prec wd. The act or process of titrating.

TITRE n See prec wd. The concentration of a solution as determined by titration.

Tl Chemical symbol for **thallium**, element at. no. 81.

Tm Chemical symbol for **thulium**, element at. no. 69.

TOLUENE n Coined fr *Tolu* (*balsam*) and suff **-ene**. The aromatic hydrocarbon $C_6H_5CH_3$, contained in Tolu balsam, which is obtained by incision in the bark of the Tolu tree (named after the town Santiago de Tolú, in Columbia).

TOLYL adj From **tolu(ene)** (q.v.) and suff **-yl**. The radical $C_6H_4CH_3$.

TORR n After *Torricelli* (1608–47). Unit of pressure formerly used in vacuum technology (= 1 mm Hg = 101325/760 Nm^{-2}).

TOURMALINE n F *tourmaline*, Ger *Turmalin*, ult derived fr Sinhalese *toramalli*, a general name for the cornelian (a semi-transparent variety of quartz). A complex aluminium borosilicate occurring naturally.

TOXIC adj L *toxicum* 'poison', fr Gk τοξικόν φάρμακον (toxicon pharmacon) 'arrow poison' fr τόξον (toxon) 'bow', and φάρμακον (pharmacon) 'drug'. Pertaining to poison; poisonous.

TOXICITY n See prec wd. Poisonous quality, esp in relation to its strength.

TOXIN n From **tox(ic)** (q.v.) and suff **-in**. A specific poison, esp one produced by a microbe, which causes a particular disease when produced in a human or animal body.

TRANS- Prefix meaning 'across, over, beyond, through'. L *trans*. (Contrast **cis-**.)

TRANSITION n L *transitus*, 'a going across or over'. A passing over from one state to another.

TRANSLATIONAL adj F *translation*, from L *translatio*, 'a carrying over'. Applied to the type of molecular motion in which the molecules move about, in contrast to rotational and vibrational motion.

TRI- Combining form meaning 'three, thrice, three-fold'. L *tri-*, Gk τρι- (tri-) 'three, threefold'.

TRIBO- Combining form meaning 'pertaining to friction'. Gk τρίβειν (tribein), 'to rub'.

TRIBOLOGY n From **tribo-** (see prec entry), and **-logy** (q.v.). The science dealing with friction and lubrication.

TRIBOLUMINESCENCE n From **tribo-** (q.v.) and *luminescence* (see **luminescent**). The evolution of (usually faint) light as a result of rubbing or grinding a solid.

TRICLINIC adj From **tri-** (q.v.) and Gk κλίνειν (kleinein) 'to bend, to cause to slope'. Having three axes intersecting at oblique angles.

TRIDYMITE n Ger *Tridymit*, fr Gk τρίδυμος (tridymos) 'three-fold', formed with τρίς (tris), 'thrice', by analogy with διδυμος (didymos), 'twin'. A mineral form of silica which commonly occurs in trillings, i.e. crystals composed of three individuals.

TRIGONAL adj From **tri-** (q.v.) and Gk γωνία (gonia) 'corner, angle'. Triangular, applied to a figure having triangular faces.

TRIS Combining form meaning 'three times', (see **tri-**). From Gk τρίς (tris) 'thrice'.

TRITIATED adj See foll wd. Containing **tritium** in place of one or more hydrogen atoms originally present.

TRITIUM n Gk τρίτος (tritos), 'third'. The third isotope of hydrogen of mass number 3 (cf. protium and deuterium), i.e. ^3_1H symbol **T**.

TRITURATE v L *trituratus*, pp of triturare, 'to thresh, grind'. To rub, or grind (as in a mortar).

TROPINE n From **atropine** (q.v.). An alkaloid, $C_8H_{15}ON$, with a 7-membered carbon ring and a nitrogen bridge, which is prepared by the hydrolysis of atropine.

TROPOLONE n Name coined by Dewar in 1945 for a compound, C_7H_6O, which contains a keto group ('-**one**') and, like **tropine** (q.v.), a seven-membered carbon ring.

TRYPSIN n Ger *Trypsin*, coined by Kühne in 1874, probably fr Gk τρύειν (tryein), 'to rub down, wear out', with the ending -*psin* by analogy with pepsin. An enzyme present in the pancreatic juice and obtained by Danilewsky in 1862 by rubbing down the pancreas with glycerin.

TRYPTOPHAN n From **tryp(sin)** (see prec wd) and Gk φαινείν (phainein), 'to cause to appear'. One of the essential amino-acids, which can be obtained from proteins by enzymic hydrolysis with trypsin.

TUNGSTEN n From Swedish *tung* 'heavy', *sten*, 'stone', name coined by Scheele in 1780 in allusion to the high density of the minerals containing the metal. The dense metal of high melting point, element at. no. 74, symbol **W** (fr Ger **Wolfram** (q.v.)).

TURPENTINE n L *terebinthus*, fr Gk τερέβινθος (terebinthos), 'the terebinth or turpentine tree'. A resin exuded from the terebinth and various other coniferous trees. Often used as abbreviation for 'turpentine oil'. (See foll wd.)

TURPENTINE OIL or **SPIRITS OF TURPENTINE** n Obtained by distillation of the resin exuded by various conifer trees. (See prec wd.)

TYROSINE n Coined by Liebig fr Gk τυρός (tyros), 'cheese', and suff -**ine** denoting an amino-acid. An amino acid, $C_9H_{11}NO_3$, which is readily obtained from cheese by fusing casein with alkali.

U

U Chemical symbol for **uranium** (q.v.) element at. no. 92.

ULTRA- Prefix meaning 'beyond'. L *ultra* 'on the other side, beyond'.

ULTRACENTRIFUGE n From **ultra-** (q.v.) and **centrifuge**. Name introduced by Svedberg and Rinde in 1924 for a centrifuge capable of running at very high speeds.

ULTRAMARINE n Med L *ultramarinus*, 'from beyond the sea'. A blue pigment which earlier had to be imported 'from beyond the seas' in the form of *lapis lazuli* (a mineral); a complex silicate of aluminium and sodium, containing sulphur.

ULTRAVIOLET n From **ultra-** and *violet*. That part of the spectrum which is immediately 'beyond' (i.e. at higher frequencies than) the violet end of the spectrum.

UN- Prefix denoting reversal, removal or deprivation.

UNDECANE n L *undecim*, 'eleven', fr L *unus*, 'one', and *decem*, 'ten', with suff **-ane**, denoting an alkane. The saturated hydrocarbon $C_{11}H_{24}$.

UNI- Combining form meaning 'one, single'. L *unus*, 'one'.

UNIMOLECULAR adj From **uni-** (q.v.) 'one', and molecular (see **molecule**). Involving only one molecule. Usually applied to reaction mechanisms and implying that the reaction proceeds by the decomposition of individual molecules, rather than by pairs of molecules (bimolecular) or by triads (termolecular).

UNIVALENT adj From **uni-** (q.v.) and L *valens*, gen *valentis*, pres p of *valere*, 'to have power'. Having a valency of one.

UNSATURATED adj From **un-** (q.v.) and **saturated** (q.v.). (i) Of a solution; having a concentration lower than the saturation concentration. (ii) Of a vapour; having a pressure lower than the saturation value. (iii) Of a compound; having the capability of undergoing addition reactions.

URANIUM n After *Uranus*, the planet discovered by Herschel in 1781; name coined, in honour of Herschel's discovery, by Klaproth in 1789 for the new element found in pitchblende. The dense metal of at. no. 92, till recent years the final element in the periodic table. Symbol **U**.

UREA n F *urée*, coined by de Fourcroy in 1799 from the base of the word **urine** (q.v.). The compound $CO(NH_2)_2$ which is found in urine.

URINE n F fr L *urina*, which is cognate with Old Icelandic *var*, *vare*, 'water', and words in other European languages. All of these derive from the Indo–European base *wer*, 'water, wet'.

UV or **uv** Contraction for **ultraviolet** (q.v.).

V

V Chemical symbol for **vanadium** (q.v.), element at. no. 23. Symbol for **volume**. Symbol for **volt** in the SI system.

v Symbol for specific volume.

VALENCE, VALENCY n L *valentia*, 'strength, capacity'. The quality which determines the number of atoms or groups with which one atom of a given element will unite.

VALERIC ACID n From L *Valeriana officinalis*, the valerian plant. Isovaleric acid, $(CH_3)_2CHCH_2COOH$ occurs in the root of the valerian.

VALINE n From **val(eric)** and suff **-ine** denoting an amino-acid. α-Aminoisovaleric acid $(CH_3)_2CH$-$CH(NH_2)COOH$. (See prec wd.)

VANADIUM n After *Vanadis*, Scandinavian goddess of beauty, name coined by Sefström in 1830 for the metal he found in a Swedish iron ore. The metallic element, at. no. 23, symbol **V**.

VANILLIN n From *vanilla*, which occurs in the vanilla pod (named, from the shape of the bean, from Sp *vainilla*, dim of *vaina*, 'sheath, pod'). The compound $C_6H_3(OH)(OCH_3)$-CHO the main chem constituent of the odoriferous material in vanilla.

VAPOUR n F *vapeur*, fr L *vapor*, 'steam, vapour'.

VERDIGRIS n OF *vert de grece*, lit 'green of Greece'. The greenish blue corrosion layer (basic copper carbonate) on copper, bronze or brass; also basic copper acetate used as a pigment.

VERMICULITE n L *vermiculus*, dim of *vermis* 'worm'. Any one of a group of micaceous minerals, (hydrous silicates), so called because when slowly heated they open into long worm-like structures.

VERONAL n Coined by its discoverer, Emil Fischer (1852–1919), from the name of the town of *Verona*, in Italy, reputedly because he was near Verona when giving a name to this substance. Diethylmalonyl urea, $CO(NHCO)_2C(C_2H_5)_2$ a hypnotic, more usually termed 'barbitone'.

VESICANT n fr L *vesica*, 'blister' and suff *-ant*. Any substance which blisters the skin.

VICINAL adj L *vicinus*, 'neighbouring'. Neighbouring, as opposed to distant. Used in particular of adjacent atoms, ions or groups in a molecule, in a solid substance or on a solid surface. Abbrev **vic-**.

VINYL n Coined from L *vinum* 'wine', and suff **-yl**. The radical C_2H_3.

VIRIAL n or adj Coined by Clausius in 1870 from the stem of L *vires*, pl of *vis*, 'strength'. Usually encountered as *virial coefficient*. In the expression $PV = RT(1 + B/V + C/V^2 + D/V^3 + ...)$, B, C, D ... are the first, second, third ... virial coefficients; they are a measure of the strength of interaction of the molecules in a gas.

VISCOSITY n Late L *viscosus*, 'full of bird-lime, sticky', fr L *viscum*, 'mistletoe, bird-lime'. The quality of being viscous, i.e. of showing a resistance to flow.

VITAMIN also **VITAMINE** n Ger *Vitamin*, name coined by C. Funk in 1913, fr L *vita*, 'life', and Ger *Amin*, **amine** (q.v.), for accessory food factors, in the (erroneous) belief that they were amines.

VITREOUS adj L *vitreus*, fr *vitrum*, 'glass'. Resembling glass, glass-like.

VITRIOL n F *vitriol*, fr Med L *vitriolum*, fr L *vitrum* 'glass'. One or other of the sulphates of a number of metals used in the arts or medicinally, especially ferrous sulphate. So called because these sulphates have a glass-like appearance.

VOLT n Named after the Italian physicist *A. Volta* (1745–1827). Unit of electrical potential difference in the SI system.

VOLTAIC adj Named after *Volta* (see prec wd) who investigated the production of electricity by chemical reaction. Pertaining to electricity produced by chemical reaction.

VOLUME n F *volume*, fr L *volumen*, 'a roll of parchment, a book', fr *volvere*, 'to roll'. Size, dimensions or bulk of a book; hence the scientific meaning of 'cubic contents' (1812).

VULCANISE v A hybrid coined by W. Brockenden in 1857, from *Vulcan*, god of fire in Roman mythology (cf. volcano) and suff -ise. The process of converting raw rubber into a strong elastic material by treatment with sulphur, often at elevated temperature.

W Chemical symbol for **tungsten** (q.v.), element at. no. 74, fr Ger **Wolfram**. Symbol for **watt**, unit of power in the SI system.

WATER n OE *waeter*, related to similar words in numerous other European languages; Hittite *watar*; Gk ὕδωρ (hydor), 'water', cf. O Slav, Russ, *voda*, 'water'.

WATT n Named after *James Watt* (1736–1819). Unit of power in the SI system. Symbol **W**.

WILLEMITE n Ger *Willemit*, named by Lévy in 1830 in honour of *Willem I*, King of the Netherlands. A mineral, zinc orthosilicate.

WITHERITE n Name given by Werner in 1790 to the mineral form of barium carbonate, in honour of *Dr. Withering* who discovered and analysed it.

WOLFRAM n *Wolfram* (old name of tungsten minerals, applied derisively in allusion to the fact that they were confused with tin ore but yielded no tin. Agricola (1556) referred to the mineral now known as **wolframite** as 'wolf foam' because it seemed to 'eat up' tin in the same way as a wolf eats up sheep).

WOLFRAMITE n See prec wd.

WOLLASTONITE n Name given by J. Lehman in 1818, to a mineral form of calcium silicate, $CaSiO_3$ after *W. H. Wollaston* (1766–1828).

X

XANTHATES n Gk ξανθός (xanthos) 'yellow'. Salts of unstable acids of the type ROC(:S)SH, (R = alkyl group). The cuprous salts are yellow, hence the name.

XANTHINE n See xanth(o)-. 2,6 dioxypurine (see **purine**), originally called xanthic acid on account of the yellow residue left when evaporated with nitric acid.

XANTHO- (before a vowel **XANTH-**) Combining form meaning 'yellow'. Gk ξανθός (xanthos), 'yellow'.

XANTHONE n From xanth(o)- (q.v.) and -one, denoting a ketone. Benzophenone oxide, $CO(C_6H_4)_2O$, which on reduction gives xanthydrol which gives yellow solutions with acids.

XANTHOPHYLLS n (pl) From xanth(o)- (q.v.) and phyll(o)- (q.v.). Yellow pigments which occur in grasses and green leaves. The name was originally given to the pigment of formula $C_{40}H_{56}O_2$.

Xe Chemical symbol for **xenon** (q.v.), element at. no. 54.

XENON n Gk ξένος (xenos) 'stranger', and -on (by analogy with argon, krypton, etc.); name coined by Ramsay and Travers in 1898 for the noble gas present in the atmosphere; element at. no. 54, symbol **Xe**.

XERO- (before a vowel **XER-**) Combining form meaning 'dry'. Gk ξηρός (xeros), 'dry'.

XEROGEL n From **xero-** (see prec wd) and **gel**. A gel from which the adherent dispersion medium has been removed by evaporation.

XYLENE n From xyl(o) 'wood' (see next wd) and suff -ene. Any of the three isomeric aromatic hydrocarbons $C_6H_4(CH_3)_2$ (ortho-, meta- and para-xylene); a mixture of these is present in the distillate from the destructive distillation of wood.

XYLO- (before a vowel **XYL-**) Combining form denoting 'wood'. Gk ξύλον (xylon), 'wood'.

XYLOSE n From xyl(o) (q.v.) and suff -ose denoting a sugar. The saccharide $HOCH_2(CHOH)_3CHO$, the dextro (+) form of which can be obtained by boiling wood-gum with dil sulphuric acid.

Y

Y Chemical symbol for **yttrium** (q.v.), element at. no. 39.

Yb Chemical symbol for **ytterbium** (q.v.), element at. no. 70.

-YL Combining form used to denote the names of radicals, e.g. ethyl, phosphoryl. Introduced into chemistry by Liebig and Wöhler in 1832 when they coined the name **benzoyl** (q.v.).

YTTERBIUM n From *Ytterby*, Sweden, place from which was obtained the mineral gadolinite in which Mosander in 1843 discovered the element erbium, from which later Marignac (1878) separated ytterbium (cf. terbium, yttrium). The rare lanthanide element, at. no. 70; symbol **Yb**.

YTTRIUM n From *Ytterby*, Sweden, place from which was obtained the mineral gadolinite from which Gadolin in 1794 and Ekeberg in 1797 separated an earth which they named yttria, and in which Mosander in 1843 discovered the element (cf. erbium, terbium, ytterbium). The metallic element of at. no. 39, found in association with the lanthanum series of elements; symbol **Y**.

Z

Z Symbol for collision number (no. of molecular collisions per unit volume per unit time).

ZEOLITE n Coined by the mineralogist A. F. Cronstedt in 1756, fr Gk $\zeta\epsilon\hat{\iota}\nu$ (zein), 'to boil', and $\lambda\iota\theta os$ (lithos), 'stone', for a mineral species which appeared to boil when heated in the blowpipe. A group of hydrated aluminosilicates principally of calcium and of sodium, which occur naturally or can be made artificially.

ZERO n F *zéro*, Ital *zero*, earlier *zefiro* (fr Med L *zephirum*) fr Arab *sifr*, 'empty, zero'.

ZETA-POTENTIAL n A term in colloid chemistry, so-called because by convention it is symbolised by the Greek letter ζ (zeta). The work required to bring unit charge from infinity to the surface of shear between a colloidal particle and the surrounding medium.

ZINC n Ger *Zink*, possibly from *Zinke* 'spike, tooth', in reference to the spiky appearance of the metal as it crystallises. The metallic element, at. no. 30; symbol **Zn**.

ZINGIBERENE n Coined fr Gk $\zeta\iota\gamma\gamma\iota\beta\epsilon\rho\iota s$ (ziggiberis), 'ginger', fr Sanskrit *śringa-vêra*, 'antler-shaped'. A sesquiterpene $C_{15}H_{24}$ which is the main constituent of ginger oil.

ZIRCONIUM n From *zircon*, the mineral zirconium silicate; name coined by Klaproth in 1789 for the new element he found in the mineral zircon. The metallic element of at. no. 40; symbol **Zr**.

Zn Chemical symbol for **zinc** (q.v.), element at. no. 30.

Zr Chenical symbol for **zirconium** (q.v.), element at. no. 40.

ZWITTERION n Ger *Zwitter*, 'hybrid, hermaphrodite', (cf. Ger *zwei*, 'two') and **ion**. An ion (usually large) which has a positive charge at one site and a negative charge at another site remote from the first.

APPENDIX 1

GREEK ALPHABET

Greek letter		Name	English equivalent
A	α	alpha	a
B	β	beta	b
Γ	γ	gamma	g (hard)
Δ	δ	delta	d
E	ϵ	epsilon	e
Z	ζ	zeta	z
H	η	eta	e
Θ	θ	theta	th
I	ι	iota	i
K	κ	kappa	k, c
Λ	λ	lambda	l
M	μ	mu	m
N	ν	nu	n
Ξ	ξ	xi	x
O	o	omicron	o
Π	π	pi	p
P	ρ	rho	r (rh)
Σ	$\sigma\ s$	sigma	s
T	τ	tau	t
Υ	υ	upsilon	u, y
Φ	ϕ	phi	ph
X	χ	chi (hard)	kh
Ψ	ψ	psi	ps
Ω	ω	omega	o

APPENDIX 2

LATIN AND GREEK NUMERALS

Number		Latin	Greek
1	One	unus	ἕν (hen)
2	Two	duo	δύο, δι- (duo, di-)
3	Three	tres	τρία (tria)
4	Four	quattuor	τέτταρες, τετρα- (tettares, tetra-)
5	Five	quinque	πέντε (pente)
6	Six	sex	ἕξ (hex)
7	Seven	septem	ἑπτά (hepta)
8	Eight	octo	ὀκτώ (octo)
9	Nine	novem	ἐννέα (ennea)
10	Ten	decem	δέκα (deca)
100	Hundred	centum	ἑκατόν (hecaton)
1 000	Thousand	mille	χίλιοι (chilioi)
1st	First	primus	πρῶτος (protos)
2nd	Second	secundus	δεύτερος (deuteros)
3rd	Third	tertius	τρίτος (tritos)

APPENDIX 3

LATIN PREFIXES

a, ab, or abs	away from	inter	between, among
ad	to	intra or intro	within, inside
(also occurs as a-,		juxta	next to
ac-, af-, al-, am-,		ob	against, opposed,
ap-, ar-, as-, at-)		(also occurs	in the way of
ante	before	as o-, oc-, of-	
circum	all around	and op-)	
cis	on this side of	per	through,
	(opposite to trans)		thoroughly
con or com	together, with	post	after
(also occurs as		prae	before
co-, col-, cor-)		pro	for, instead of, forth
contra	against	re-	again; turning back;
de	down (down from		restoring to
	or away from)		original condition;
dis-	apart or asunder		intensification
e or ex	out of	retro	behind, backward
(also as ef-, ec-)		se-	apart, without
extra	outside	sub	under
in	into or on	(also occurs as	
(when before a		suc-, suf-, sup-)	
verb)		super	above, over
(also occurs as		trans	across, over,
il-, im-, ir-)			beyond, through
in	not	ultra	beyond
(when before an			
adjective)			
(also occurs as			
il-, ig-, im-, ir-)			

APPENDIX 4

GREEK PREFIXES

a-, an-	ἀ- ἀν-	without	hypo	ὑπό	below, under
amphi	ἀμφί	both, around	cata	κατά	down, against
ana	ἀνά	up	meta	μετά	after, change
anti	ἀντί	against	ortho-	ὀρθο-	straight, right
apo	ἀπό	off, away from	para	παρά	beside, similar,
dia, di-	διά δι-	through, asunder			near, beyond,
en	ἐν	in, on			irregular, a
(also occurs as el- or em-)					modification of
endo-	ἐνδο-	within	peri	περί	around, about
epi	ἐπί	upon	syn	σύν	together, alike
ek, ex	ἐκ, ἐξ	out of	(also occurs as syl-, sym-, sys-)		
exo	ἔξω	outside			
hyper	ὑπέρ	above, over, excess			

134

APPENDIX 5

LATIN ROOTS

a, ab, abs	away from	fero	bear, carry
acidus	sharp	flavus	golden yellow
ad	towards, to	fluo	flow
aer	air	gelo	freeze
aestimo	appraise, value	geminus	twin
aether	upper air	ignis	fire
ago	drive, set in motion	laevus	left hand
albus	white	lamina	thin plate
alga	seaweed	lapis	stone
ambio	surround	latex	liquid
ante	before	libro	balance
apex	top	lignum	wood
aqua	water	ligo	bind, tie
caedo	kill	liquidus	fluid
calculus	pebble	metallum	metal
calor	heat	membrana	membrane
calx	chalk, limestone	mille	thousand
capillis	hair	misceo	mix
carbo (gen. carbonis)	burnt wood, charcoal	moles	lump, mass
		multus	many
catena	chain	nucleus	nut, kernel
centum	hundred	oleum	oil
cera	wax	orbis	circle
cis	on this side of (opp to 'trans')	pars	part
		per	through
complexus	embrace	post	after
contra	opposite, against	prae	before
corpus (gen. corporis)	body	praecipito	cast down
		proximitas	nearness
corpusculum	little body	purpureus	purple
crystallus	crystal	putresco	decay
de	down	pyramis	pyramid
decem	ten	radix	root
dexter	right hand	rubidus	reddish
dis-	apart, separately	sal	salt
duo	two	scientia	knowledge
extra	outside	semis	half

septem	seven	sorbeo	suck in
septum	hedge, wall	sub	under
sex	six	sulfur	sulphur
silex (gen	stone, flint	super	above, over
silicis)		vapor	vapour
solidus	solid, firm	vicinus	near
solvo	untie, loosen	vitrum	glass

APPENDIX 6

GREEK ROOTS

ἀ-	a-	without	ἶρις	iris	rainbow
ἀήρ	aer	air	ἴσος	isos	equal
αἰθήρ	aither	sky	κατά	kata	down
ἄλλος	allos	other	κέρας	keras	horn
ἅλς, ἅλος	hals, halos	salt	κηρός	keros	wax
ἄμφω	ampho	both	κινεῖν	kinein	to move
ἄνθραξ	anthrax	coal	κόλλα	kolla	glue
ἀντί	anti	against	κρυός	kryos	cold
ἀτμός	atmos	vapour	κρύσταλλος		
αὐτός	autos	self		crystallos	ice, crystal
βαρύς	barys	heavy	κυάνος	cyanos	blue
βίος	bios	life	κύβος	cubos	cube
γένισις	genesis	beginning, origin	κύκλος	cyclos	circle
			λευκός	leukos	white
γῆ	ge	earth	λίθος	lithos	stone
γλυκύς	glycys	sweet	λίπος	lipos	fat
γράμμα	gramma	that which is written	λύσις	lysis	loosening
			μακρός	makros	long, big
γράφειν	graphein	to write	μέγας	megas	big
γωνία	gonia	angle	μέλας	melas	black
διπλόος	diploos	double	μέρος	meros	part
δύναμις	dynamis	power	μέτρον	metron	measure
ἕλιξ	helix	spiral	μικρός	mikros	little
ἐναντίος	enantios	opposite	μόνος	monos	single
ἔργον	ergon	work	μορφή	morphe	form
ἕτερος	heteros	different	νέος	neos	new
εὐ	eu	well	ξανθός	xanthos	yellow
εὑρίσκω	heurisko	to discover	ξερός	xeros	dry
ζύμη	zyme	yeast	ξύλον	xylon	wood
ζώνη	zone	belt	ὀλίγος	oligos	few
ζῷον	zoon	animal	ὁμός	homos	same
ἥλιος	helios	sun	ὀξύς	oxus	sharp, acid
ἥμι-	hemi-	half	ὄργανον	organon	instrument
θεῖον	theion	sulphur	οὖρον	ouron	urine
θεραπεία	therapeia	care, healing	πλάσσειν	plassein	to mould
θερμός	thermos	hot	πορφύρα	porphyra	purple
θέσις	thesis	proposition	πρᾶξις	praxis	action

πῦρ	pyr	fire	φλόξ	phlox,		flame
ῥέος	rheos	stream, current	φλογ-	phlog-		
ῥόδον	rhodon	rose	φύλλον	phyllon		leaf
σάκχαρον	saccharon	sugar	φύσις	physis		nature
σκοπεῖν	skopein	to examine, look at	φωνή	phone		sound
			φώς	phos,		light
			φωτός	photos		
στερεός	stereos	solid	χίλιοι	chilioi		thousand
τέχνη	techne	art	χλωρός	chloros		green
τῆλε	tele	far	χρυσός	chrysos		gold
τροπή	trope	a turning, change	χρῶμα	chroma		colour
			χῶρος	choros		space
φορέω	phoreo	I carry	ψεύδω	pseudo		I deceive
φιλέω	phileo	I love				

138

APPENDIX 7

FORMATION OF PLURALS

Latin

aquarium, -a
bacterium
corrigendum
datum
maximum
medium
memorandum
minimum
quantum
septum
spectrum
stratum

alumnus, -i
calculus
nucleus
radius

alga, -ae
lacuna
lamella
lamina

apex, apices
calx, calces
helix, helices
latex, latices
matrix, matrices
vertex, vertices

Greek

criterion, -a
dodecahedron
icosahedron
octahedron
phenomenon
polyhedron
tetrahedron

analysis, -es
diagnosis
dialysis
electrolysis
hydrolysis
hypothesis
prosthesis
pyrolysis
synthesis
thesis
prognosis

APPENDIX 8

SI PREFIXES

(Système Internationale d'Unités)

Prefix	Symbol	Fraction	Prefix	Symbol	Multiple
deci	d	10^{-1}	deca	da	10
centi	c	10^{-2}	hecto	h	10^2
milli	m	10^{-3}	kilo	k	10^3
micro	μ	10^{-6}	mega	M	10^6
nano	n	10^{-9}	giga	G	10^9
pico	p	10^{-12}	tera	T	10^{12}
femto	f	10^{-15}			
atto	a	10^{-18}			